Published by: AoPS Incorporated
                  10865 Rancho Bernardo Rd Ste 100
                  San Diego, CA 92127-2102
                  info@BeastAcademy.com

ISBN: 978-1-934124-63-5

Written by Jason Batterson, Shannon Rogers, and Kyle Guillet
Book Design by Lisa T. Phan
Illustrations by Erich Owen
Grayscales by Greta Selman

Visit the Beast Academy website at www.BeastAcademy.com.
Visit the Art of Problem Solving website at www.artofproblemsolving.com.
Printed in the United States of America.
First Printing 2016.

Contents:

This is Practice Book 5B in a four-book series.

5A
• 3D Solids
• Integers
• Expressions & Equations

5B
• Statistics
• Factors & Multiples
• Fractions

5C
• Sequences
• Ratios & Rates
• Decimals

5D
• Percents
• Square Roots
• Exponents

For more resources and information, visit BeastAcademy.com.

This is Beast Academy Practice Book 5B.

Each chapter of this Practice book corresponds to a chapter from Beast Academy Guide 5B.

MATH PRACTICE 5B

MATH GUIDE 5B

The first page of each chapter includes a recommended sequence for the Guide and Practice books.

You may also read the entire chapter in the Guide before beginning the Practice chapter.

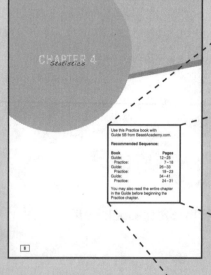

CHAPTER 4
Statistics

Use this Practice book with Guide 5B from BeastAcademy.com.

Recommended Sequence:

Book	Pages
Guide:	12–25
Practice:	7–18
Guide:	26–33
Practice:	19–23
Guide:	34–41
Practice:	24–31

You may also read the entire chapter in the Guide before beginning the Practice chapter.

Some problems in this book are very challenging. These problems are marked with a ★. The hardest problems have two stars!

Every problem marked with a ★ has a *hint!*

Hints for the starred problems begin on page 98.

Other problems are marked with a ✎. For these problems, you should write an explanation for your answer.

54.
★

55.
✎

42   Guide Pages: 39-43

Some pages direct you to related pages from the Guide.

None of the problems in this book require the use of a calculator.

Solutions are in the back, starting on page 102.

A complete explanation is given for every problem!

CHAPTER 4
Statistics

Use this Practice book with
Guide 5B from BeastAcademy.com.

Recommended Sequence:

Book	Pages:
Guide:	12-25
Practice:	7-18
Guide:	26-33
Practice:	19-23
Guide:	34-41
Practice:	24-31

You may also read the entire chapter
in the Guide before beginning the
Practice chapter.

Statistics is the study of data!

PRACTICE Ask 7 people their age, their height, how many hours of sleep they get every night, and to pick an integer from 1 to 10. Record the data you collect in the table below and answer the questions that follow.

If you can't talk to 7 people right away, you may continue working through this chapter before completing the table and questions below.

1.

Name	Age (years)	Height (inches)	Sleep per night (hrs)	Pick an integer (1-10)

All number entries should be rounded to nearest positive integer. (No fractions or decimals.)

a. If you arrange the 7 people in the table above from youngest to oldest, what will be the age of the person in the middle?

a. _____

b. What is the difference in height between the tallest person and the shortest person in the table above?

b. _____

c. What is the most common "Pick an integer" number chosen among the people you surveyed?
(If more than one number was chosen the most, list all numbers that were chosen the most. If all digits were chosen only once, write "none.")

c. _____

d. What is the sum of the number of hours slept each night by all 7 people you surveyed?

d. _____

e. If the total number of sleep hours were shared equally between all the people surveyed, how many hours would each person get?

e. _____

The middle number of an ordered list is the *median*. If there is an even number of items in the list, the median is the number halfway between the two middle numbers.

To find the median of an unordered list, we first put the numbers in order.

EXAMPLE | What is the median of the eight numbers listed below?

70 19 17 29 33 55 54 41

To find the middle number, we order these numbers from least to greatest:

17 19 29 33 41 54 55 70

The two numbers in the middle of this list are 33 and 41.
The number halfway between 33 and 41 is **37**.

Review pages 14-19 in the Guide for ways to find the number halfway between two numbers.

PRACTICE | Use the data below to answer the questions that follow. A dash in the Horn Length category means that the monster does not have horns.

Name	Age (years)	Weight (pounds)	Height (inches)	Horns (inches)	# of Legs	# of Eyes
Sam	8	55	34	7	2	2
Millie	8	39	50	11	2	3
Bortle	9	7	18	-	3	2
Snargle	8	29	28	15	4	3
Bill	8	55	48	2	2	2
Oksbert	9	70	30	-	2	2
Val	8	53	45	12	2	5
Lobwart	9	19	32	5	4	2

2. What is the median age of these monsters?

2. _____

3. What is the median weight of the *two-legged* monsters?

3. _____

4. What is the median horn length for the monsters who have horns?

4. _____

Name	Age (years)	Weight (pounds)	Height (inches)	Horns (inches)	# of Legs	# of Eyes
Sam	8	55	34	7	2	2
Millie	8	39	50	11	2	3
Bortle	9	7	18	-	3	2
Snargle	8	29	28	15	4	3
Bill	8	55	48	2	2	2
Oksbert	9	70	30	-	2	2
Val	8	53	45	12	2	5
Lobwart	9	19	32	5	4	2

PRACTICE | Use the data above to answer the questions that follow. A dash in the Horn Length category means that the monster does not have horns.

5. How many inches greater is the median height of the 8-year-old monsters than the median height of the 9-year-old monsters?

5. _____

6. Which monster has the median number of legs and eyes but is **not** the median age?

6. _____

7. Circle the monster listed to the right who could join the group above without changing the median height **or** the median weight.

Name	Height	Weight
Yosh	33 in	41 lbs
Tildie	32 in	46 lbs
Kres	34 in	41 lbs
Mitch	33 in	46 lbs
Cad	34 in	39 lbs
Erma	32 in	39 lbs

8. One more monster joins the 8-monster class shown at the top of the page. What are the possible median heights for the new group of 9 monsters?

In a **Median Split** puzzle, the goal is to draw a vertical or horizontal line through a group of numbers to separate them into two groups that have the same median.

EXAMPLE | Solve the Median Split puzzle on the right.

```
              2
        8   3   19
      1   5   4   7   6
```

There are six ways to split the group of numbers with a vertical or horizontal line, as shown below.

```
  |   2              2 |              2              2        |
  | 8  3  19       8 | 3  19        8  3 |19        8  3  19 |
 1| 5  4  7  6    1 | 5  4  7  6   1  5  4 |7  6     1  5  4  7 |6
```

```
      2                    2
   _____            
   8  3  19             8  3  19
  1  5  4  7  6      _____
                     1  5  4  7  6
```

Only the vertical bar shown below separates the numbers into two groups with the same median: 5.

```
           | 2
       8   | 3   19
     1   5 | 4   7   6
   Median: 5   Median: 5
```

PRACTICE | Solve each Median Split puzzle below.

9.   1 3 5 7
         2 4 6

10. 9 7 10
            3 4

11.   4
             2 3 6
           7 5 8

12. 1 8 4 6
         5 3 9

13.   2 6
          11 3 5
              3

14. 1 2 4
         9 5 5
         4 3

PRACTICE | Answer each question below.

15. Compute the median of **all ten numbers** in each Median Split puzzle below.

a.
```
        8
    1 4 5 9
      7 7 2
      5 3
```
Median: _____

b.
```
    2 7
    8 4 5
      4 8 7
      3 9
```
Median: _____

c.
```
      6 4 5
    3 8 8
      1 3 2
          1
```
Median: _____

16. Now, solve each of the Median Split puzzles.

a.
```
        8
    1 4 5 9
      7 7 2
      5 3
```

b.
```
    2 7
    8 4 5
      4 8 7
      3 9
```

c.
```
      6 4 5
    3 8 8
      1 3 2
          1
```

17. Compare the medians from problem 15 to the medians of the divided groups in problem 16. For each puzzle, are the medians of the ten numbers smaller, larger, or the same as the median of each group in the corresponding puzzles in problem 16?

18. Mr. G. divides his class into two teams. On each team, the median height of the monsters is 46 inches. Can we determine the median height of all the monsters in Mr. G.'s class? If so, what is it? If not, why not?

19. Is it possible for a group of numbers with a median of 10 to be split into two groups so that one group has a median of 10, but the other group has a **different** median? If so, give an example. If not, explain why not.

Adding all of the numbers in a list and then dividing by the number of numbers gives us their *average.*

The average is sometimes called the *mean.*

EXAMPLE

Phyllis fills five cartons with fresh figs. The cartons contain 19, 21, 24, 24, and 27 figs. If Phyllis instead placed all of these figs so that each carton had the same amount, how many would each carton hold?

There are a total of $19+21+24+24+27=115$ figs. Dividing these equally among 5 cartons gives $115\div5=23$ figs in each carton.

23 is called the *average* of 19, 21, 24, 24, and 27.

PRACTICE | Answer each question below.

20. Around the campfire at this year's Beast Academy retreat, Winnie ate 6 roasted marshmallows, Lizzie ate 7, Alex ate 10, and Grogg ate 33. If all of the marshmallows were instead divided equally among the four monsters, how many would each monster get?

20. _____

21. Alice's doll collection is organized on 7 shelves. Her shelves currently have 9, 13, 15, 17, 16, 11, and 24 dolls. If Alice instead divides her dolls equally among the shelves, how many dolls will be on each shelf?

21. _____

22. Six monsters work together to paint a house. Borf is paid $32, and the other five monsters are paid $20 each. The six monsters decide that everyone should get an equal share. Borf gives the other monsters some money so that all six monsters have the same amount. After that happens, how much money does each monster have?

22. _____

23. Mrs. Ryan's classroom has eight pencil cups. The cups have 7, 3, 14, 11, 18, 2, 1, and 0 pencils in them. If Mrs. Ryan instead places the same number of pencils in each cup, how many pencils will the cup with 3 pencils gain?

23. _____

EXAMPLE | Find the average of the numbers below.

1, 1, 2, 3, 10, 11, 14

To find the average of a list of numbers, we compute their sum and divide by the number of numbers.

The sum of the numbers is $1+1+2+3+10+11+14 = 42$.

There are 7 numbers in this list, so we divide their sum by 7 to compute their average.

The average is $\frac{1+1+2+3+10+11+14}{7} = \frac{42}{7} = \textbf{6}$.

PRACTICE | Compute the average of each list of numbers below.

24. 2, 3, 5, 7, 8, 11, 13

25. 5, 10, 20, 20, 25, 35, 50, 75

24. _____

25. _____

26. -20, -15, -13, -9, -8, -6, -6

27. 9.1, 9.6, 9.9, 10.5, 10.9

26. _____

27. _____

PRACTICE | Answer each question below.

28. The average of six numbers in a list is 13. If the number 2 is added to the list, what will be the average of the seven numbers in the new list? Express your answer as a mixed number in simplest form.

28. _____

29. The average of nine numbers in a list is 10. If the number 2 is removed from the list, what will be the average of the eight remaining numbers?

29. _____

EXAMPLE The average of the five numbers below is 80. Find the missing value in the list.

$$74 \quad 75 \quad 78 \quad 81 \quad \underline{\quad}$$

Five numbers with an average of 80 have a sum of $5 \cdot 80 = 400$. The sum of the four numbers we know is $74+75+78+81 = 308$. So, the missing number is $400-308 = \mathbf{92}$.

— *or* —

We consider how the numbers "balance" around the average.

74, 75, and 78 are below the average by a total of 13, and 81 is above the average by 1. So, the four given numbers are a total of 12 below the average.

$$\overset{-12}{\overline{\overset{-6 \quad -5 \quad -2 \quad +1}{74 \quad 75 \quad 78 \quad 81}}}$$

To balance all of the below-average numbers, we need a number that is above the average by 12. So, the missing number is $80+12 = \mathbf{92}$.

$$\overset{-6 \quad -5 \quad -2 \quad +1 \quad +12}{74 \quad 75 \quad 78 \quad 81 \quad \mathbf{\underline{92}}}$$

PRACTICE Fill in each missing number so that every list has the average given.

30. Average: 40

39 35 44 _____

31. Average: 20

19 14 25 18 _____

32. Average: 54

55 50 56 60 _____

33. Average: 105

111 101 103 110 _____

34. Average: 91

95 88 86 97 89 _____

35. Average: 176

174 176 180 175 170 _____

PRACTICE | Answer each question below.

36. In the first five basketball games of the season, Orange Academy scored 33, 38, 32, 30, and 40 points. Their sixth game brought their average score down to 34 points. How many points did Orange Academy score in their sixth game?

36. _____

37. Below are Teddy's scores on his math tests so far this year. What score does Teddy need on his next test so that his test average is 90?

81, 96, 100, 88, 91, 78

37. _____

38. ★ Which *two* numbers can be removed from the list below so that the average of the four remaining numbers is 78?

71, 73, 74, 78, 81, 84

38. ____ and ____

39. ★ Find the average of the four numbers below *without* computing their sum.

8,953   8,952   8,950   8,957

39. _____

40. ★ During the last Beastball season, Kat scored 5 points below the team average, Matt scored 8 points above the team average, and Pat scored 82 points. The average of Kat's, Matt's, and Pat's scores is the same as the team average. What is the team average?

40. _____

In an **Averatile** puzzle, every shape must be filled with a *positive digit* according to two rules:

1. The number in each square is the average of all the numbers in the triangles that the square shares a side with.

2. No two shapes that share a side may contain the same digit.

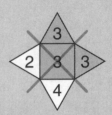

EXAMPLE | Solve the Averatile puzzle to the right.

Since each shape must be filled with a positive digit, we know that the sum of the four numbers around the square is divisible by 4. The sum of the known digits is $2+3+3 = 8$. The only digits that could be added to 8 to get a multiple of 4 are 4 and 8.

However, if we fill the triangle with a 4, the average of the four triangles around the square is $\frac{2+3+3+4}{4} = \frac{12}{4} = 3$. Adjacent shapes may not contain the same digit, so we cannot place a 3 in the square.

If we place an 8 in the empty triangle, the average of the four triangles is $\frac{2+3+3+8}{4} = \frac{16}{4} = 4$. This is the only solution.

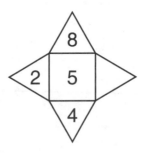

PRACTICE | Solve each Averatile puzzle below.

41.

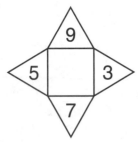

42.

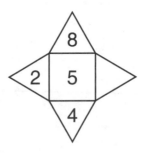

43.

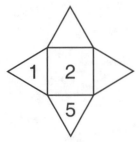

44.

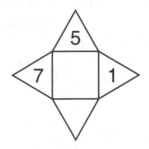

PRACTICE | Solve each Averatile puzzle below.

45.

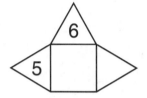

46.

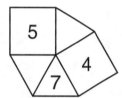

47.

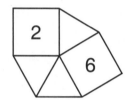

48.

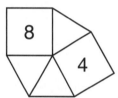

49.

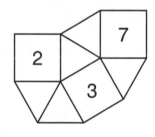

50.

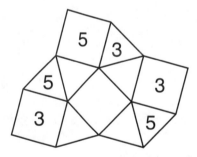

51.

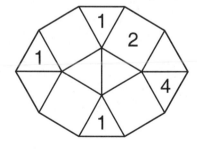

52.

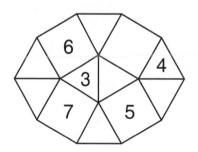

PRACTICE | Solve each Averatile puzzle below.

53.
★

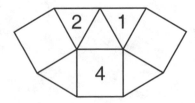

54.
★

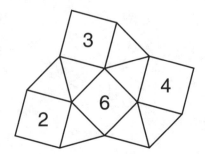

55.
★

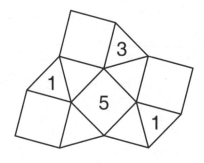

56.
★

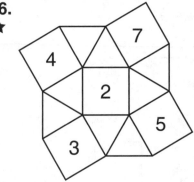

57.
★
★

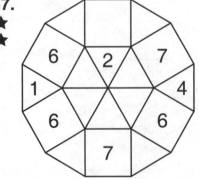

58.
★
★

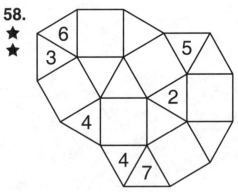

Print more Averatile puzzles at BeastAcademy.com.

PRACTICE | Answer each question below.

59. Compute the average of the five numbers below.

1   2   3   4   1,000

59. _____

60. What is the average of nineteen 2's and one 182?

60. _____

61. At Beast Burger, the manager earns $45 per hour. The other four employees each earn $10 per hour. What is the average hourly wage for an employee at Beast Burger?

61. _____

62. Grogg received 100 points on each of the first 4 homework assignments he turned in this school year. He lost his fifth assignment and received 0 points. What is Grogg's homework average for the first five assignments?

62. _____

63. In each of the sets of data above, one value is very different from the rest. Consider the median and average of each data set with and without that extreme value. Which is more impacted by one extreme value, the average or median? Explain.

EXAMPLE

Julius has eight pets: 3 dogs and 5 cats.
The average weight of his dogs is 39 pounds, and
the average weight of his cats is 7 pounds.
What is the average weight of Julius's eight pets?

Three dogs whose average weight is 39 pounds weigh
a total of $3 \cdot 39 = 117$ pounds. Five cats whose average
weight is 7 pounds weigh a total of $5 \cdot 7 = 35$ pounds.

So, the weight of all eight pets is $117 + 35 = 152$ pounds.

The average weight of all eight pets is therefore
$\frac{152}{8} = \mathbf{19\ pounds}$.

PRACTICE | Answer each question below.

64. Monster Milkshakes serves an average of 78 customer per day
Monday through Friday and an average of 120 customers per day
Saturday and Sunday. What is the average number of customers
served per day during a week at Monster Milkshakes?

64. _____

65. There are 12 kangaroosters in a field. The 8 adult
kangaroosters can leap an average of 30 feet. The 4 juvenile
kangaroosters can leap an average of 21 feet. What is the
average leaping distance among these 12 kangaroosters?

65. _____

66. There are 10 Beastball players and 40 members of the school band
riding the bus to a game. The average heights of the two groups are
different. Is the average height of all 50 students on the bus closer to
the average height of the Beastball players, or to the average height
of the band members? Explain.

PRACTICE | Answer each question below.

67. Maggie scored an average of 8 points per round in the first three rounds of an archery tournament. By the end of ten rounds, she improved her tournament average to 15 points per round. How many points per round did Maggie average in the last seven rounds of the tournament?

67. _____

68. The average weight of the six members of the Teenie Wrestling Team is 15 pounds. When Jake, Ann, and Ungor join the team, the average weight decreases to 12 pounds. What is the average weight of Jake, Ann, and Ungor?

68. _____

69. The average value of the 40 hockey cards in Dave's collection is $6. After receiving 8 new cards for his birthday, the average value of a card in his collection increases to $8. What is the average value of the cards Dave received on his birthday?

69. _____

70. Jack has 15 gold coins worth an average of 6 dollars, 33 silver coins worth an average of 6 dollars, and 57 bronze coins worth an average of 6 dollars. What is the average value of all of Jack's coins?

70. _____

71. Ms. Maple's tulips have an average height of 25 cm. Her daffodils have an average height of 5 cm. What is the average height of the 222 tulips and 333 daffodils in Ms. Maple's garden?

71. _____

EXAMPLE

In a game of ring toss, each toss is worth either 4 points or 9 points. Raven averages 7 points per toss in a single game of ring toss. If she made ten 4-point tosses, how many 9-point tosses did she make?

Method 1: Balance around the average.

We compare the score of each toss to Raven's average score. Each 4-point toss is three points *below* Raven's average, and each 9-point toss is two points *above* her average.

Raven made ten 4-point tosses for a total of $10 \cdot 3 = 30$ points below her average.

To balance this, she needs enough 9-point tosses to equal 30 points above her average.

$$
\begin{array}{cc}
\overbrace{\quad -30 \quad} & \overbrace{\quad +30 \quad} \\
{\scriptstyle -3\ -3\ -3\ -3\ -3\ -3\ -3\ -3\ -3\ -3} & {\scriptstyle +2\ \cdots\ +2} \\
4\ 4\ 4\ 4\ 4\ 4\ 4\ 4\ 4\ 4 & 9\ \cdots\ 9
\end{array}
$$

Each 9-point toss is 2 points above her average, so she made $\frac{30}{2} =$ **15** nine-point tosses.

Method 2: Write and solve an equation.

We write two different expressions for the number of points Raven scored.
Let n represent the number of 9-point tosses she made.

- Raven scored $9n$ points with n nine-point tosses and $4 \cdot 10 = 40$ points with ten 4-point tosses for a total of $9n + 40$ points.

- Raven made $n + 10$ tosses with an average score of 7, so she scored a total of $7(n + 10)$ points.

We have two expressions for the number of points Raven scored, so we can write an equation: $9n + 40 = 7(n + 10)$.

We distribute the 7 to get $9n + 40 = 7n + 70$. Solving for n, we get $n = 15$.

So, Raven made **15** nine-point shots.

> Review solving equations in Chapter 3 of Beast Academy 5A!

PRACTICE | Answer each question below.

72. Ernie scores 0 points during his first Rugball game and 60 points in every game after that.

a. What is Ernie's average score after 2 games?

a. _____

b. What is his average score after 5 games?

b. _____

c. After how many total games will Ernie's average score be 54?

c. _____

d. After how many total games will Ernie's average score be 59?

d. _____

PRACTICE | Answer each question below.

73. A bag of coins containing only nickels and quarters has 12 nickels. If the average value of a coin in the bag is 9 cents, how many quarters are in the bag?

73. _____

74. How many -2's must be added to a list of eight 6's so that the average of all the numbers in the list is 0?

74. _____

75. In Professor Wolfe's class, homework assignments are worth 100 points if turned in and complete. Incomplete assignments are worth 70 points. Penelope turned in all of her homework assignments, but 12 of her assignments were incomplete. Her homework average is 92. How many assignments did Penelope turn in?

75. _____

76. ★ On Tuesday morning, the average weight of a flamingoat at the shelter is 8 pounds. After Trina adopts a 28-pound flamingoat from the shelter, the average weight of the remaining flamingoats at the shelter is 3 pounds. How many flamingoats were at the shelter on Tuesday morning?

76. _____

77. ★ The average age of all the monsters at a family reunion is 36. After 36-year-old Cousin Betsy arrives with her 11-year-old son, the average age of all the monsters is 35. How many monsters are now at the reunion, including Betsy and her son?

77. _____

Average and median don't always give us a good picture of our data. Consider the two lists of data below:

$$5, 5, 5 \qquad -60, -12, 0, 1, 9, 19, 33, 50$$

These data sets are very different, but they share the same median and average (5). Two other statistics that help us note differences are the mode and range.

The **mode** is the number that occurs most often in a list. A list can have more than one mode. If every number in a list occurs exactly one time, then we say that the list has **no mode**.

The **range** is the difference between the smallest and largest numbers in a data set.

We can also use statistics like **mode** and **range** to describe our data.
Range tells us how spread out our data is.

The group (5, 5, 5) has a mode of 5 and a range of 0.

The group (-60, -12, 0, 1, 9, 19, 33, 50) has no mode and a range of 110.

PRACTICE | Determine the mode or modes for each list of numbers below. If a list has no mode, write "no mode."

78. 5, 10, 20, 20, 25, 35

79. 2, 2, 3, 4, 6, 6, 13

78. _____

79. _____

80. -3, -4, -4, 4, 5, 5, 5

81. 9.6, 9.9, 10.1, 10.5, 10.9

80. _____

81. _____

PRACTICE | Determine the range for each list of numbers below.

82. -32, 24, 35, 7, 48, -16

83. 2.5, 1.2, 3.9, 4.0, 2.8

82. _____

83. _____

84. A list of three monsters' ages has a mode of 6 and a range of 7. How old is the oldest monster?

84. _____

PRACTICE | Answer each question below.

85. What is the smallest possible range for a list of three integers with a mean of 6, a median of 7, and no mode?

85. _____

86. ★ The mean, median, and mode of this list are all the same number. What is x?

$$100, 500, 600, 400, x$$

86. _____

87. ★ A group of 13 **different** integers has a median of 15 and a range of 17. What is the greatest possible number that could be in this group?

87. _____

88. ★ The heights in centimeters of 7 flowers are distinct positive integers with a mean of 80 and a median of 81. What is the greatest possible range of the flower heights?

88. _____

89. ★ In a class of ten monsters, removing the tallest monster decreases the average height by 6 inches. Removing the shortest monster increases the average height from 28 to 30 inches. What is the range of the monster heights?

89. _____

In a **Stat Square** puzzle, each of the nine white squares is filled with a *positive digit* so that the clues given in the surrounding shaded squares give the correct average, median, mode, and range for the row or column that they label.

For example, the Stat Square below can be filled as shown so that the range of the top row is 3, the average of the middle column is 5, the mode of the middle row is 1, and the median of the right column is 4.

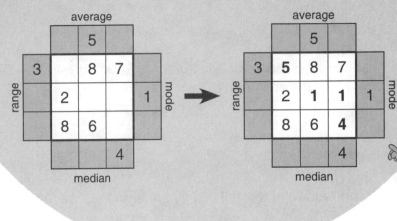

You don't need to fill in the remaining blank squares on the outside!

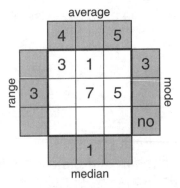

PRACTICE | Solve each Stat Square puzzle below. A mode clue of "no" means that the row has no mode.

90.

average: 5
range: 4
row: 1 3
row: 5 7 2
row: 4
mode: 4
median

91.

average: 4 5
row: 3 1
mode: 3
range: 3
row: 7 5
mode: no
median: 1

92.

average: 7
row: 3
mode: 1
range: 2
row: 3 2
median: 3, 2

93.

average: 6
row: 5 8, mode: no
mode: 2
range: 6
row: 8 4
median: 5

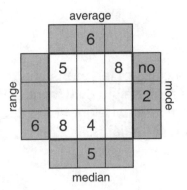

PRACTICE | Solve each Stat Square puzzle below. A mode clue of "no" means that the row has no mode.

94.

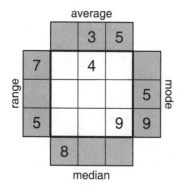

95.

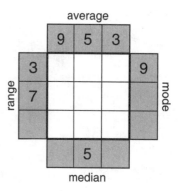

96. ★

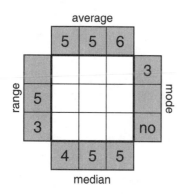

97. ★

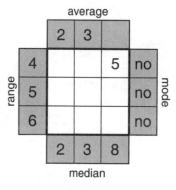

98. ★★

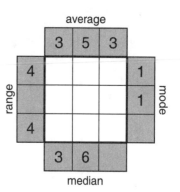

99. ★★

A *histogram* groups numerical data into equally-spaced ranges on a number line.

The heights of the 10-year-old monsters at Beast Academy were measured to the nearest inch. The histogram below displays the number of monsters that fall into each range of heights.

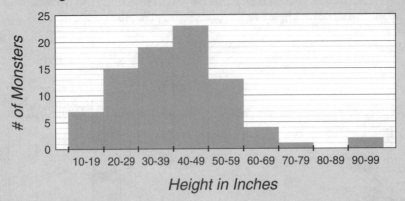

In this histogram, we see that there are 7 monsters who are 10 to 19 inches tall and 15 monsters who are 20 to 29 inches tall.

PRACTICE | Use the histogram above to answer the questions that follow.

100. Which of the nine height ranges above contains the greatest number of monsters?

100. _____

101. How many of the monsters are at least 70 inches tall?

101. _____

102. How many monsters' heights are included in the graph above?

102. _____

103. Circle the value below that *could* be the median height of these monsters.

15 in   22 in   29 in   36 in   43 in   50 in   57 in   64 in

104. Use the data in the histogram above to create a new histogram on the right.

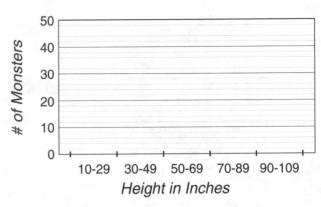

A *line graph* can be helpful for displaying and comparing changes in data over time.

The line graph below displays the scores of four students on a monthly test from September to February.

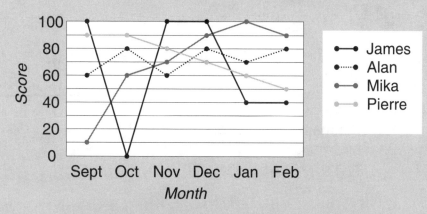

From this line graph, we see that James scored 100 points on his September test.

PRACTICE | Use the line graph above to help you answer the questions that follow.

105. Which student has the highest average score?   105. _____

106. Which student has the highest median score?   106. _____

107. Which student has the highest mode score?   107. _____

108. Which student's scores have the smallest range?   108. _____

109. Whose scores have improved the most overall in the past 6 months? 109. _____

110. Based on this graph, who would you want to study with for the March test?
✎   Explain.

111. Mr. Erikson looks at the scores of three of the four students above.   111. _____
★   If he determines the mode score is 80, whose scores did Mr.
     Erikson not include?

112. Use the data in the line graph
     above to plot the median score for
     each monthy test.

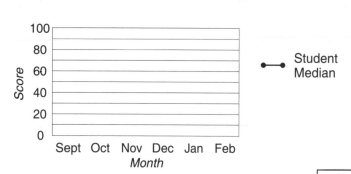

PRACTICE | Answer each question below.

113. The median of a list of consecutive integers is s. Write an expression for the average of this list.

113. _____

114. ★ We can arrange the digits 1, 2, and 3 to form six different 3-digit numbers. What is the average of these six numbers? *As an extra challenge, try to compute this average without adding up all six numbers.*

114. _____

115. ★ Six islands are connected by bridges. Each bridge connects one pair of islands. An average of 3 bridges touch each island. How many bridges are there?

115. _____

116. ★ The average of x, y, and z is 6. What is the average of $3x+y$, $3y+z$, and $3z+x$?

116. _____

PRACTICE | Answer each question below.

117. A list of consecutive integers has an average of $10\frac{1}{2}$ and ★ a range of 31. What is the smallest integer in this list?

117. _____

118. For what values of n are the median and the average of the ★ list below equal? ★

17, 12, 21, 30, n

118. _____

119. For what values of a in the list below are the mean, ★ median, and mode consecutive integers (in any order)? ★

5, 5, 7, 8, 13, a

119. _____

120. Paulie has a list of 7 integers. The mean is 18, the median is 20, the ★ mode is 21, and the range is 7. Write Paulie's list below. ★

___ , ___ , ___ , ___ , ___ , ___ , ___

CHAPTER 5
Factors & Multiples

Use this Practice book with
Guide 5B from BeastAcademy.com.

Recommended Sequence:

Book	Pages:
Guide:	42-57
Practice:	33-45
Guide:	58-66
Practice:	46-55
Guide:	67-75
Practice:	56-63

You may also read the entire chapter
in the Guide before beginning the
Practice chapter.

The **factors** of n are the integers that n is divisible by.

In this chapter, when we talk about the factors of a number, we mean the number's **positive** factors.

A **prime** number has only two factors: 1 and itself.

A **composite** number has more than two factors.

EXAMPLE | List the factors of 84.

We write the factors of 84 in pairs, starting with 1 · 84:

$$\underline{84}$$
1 · 84
2 · 42
3 · 28
4 · 21
6 · 14
7 · 12

So, the factors of 84 are
1, 2, 3, 4, 6, 7, 12, 14, 21, 28, 42, and **84**.

PRACTICE | Answer each question below.

1. List the factors of each number below from least to greatest:

18: _____   48: _____

37: _____   54: _____

49: _____   225: _____

2. What is the largest prime factor of 140?   2. _____

3. How many factors of 80 are also factors of 56?   3. _____

EXAMPLE | Write the prime factorization of 198.

To find the prime factorization of 198, we create a factor tree. We circle factors that are prime.

Review factor trees in Chapter 7 of *Beast Academy 4C*.

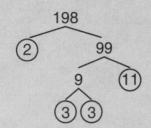

The prime factorization of a prime number is just the number itself.

When writing a number's prime factorization, we order its primes from least to greatest and use exponents for repeated factors.

So, the prime factorization of 198 is **$2 \cdot 3^2 \cdot 11$**.

PRACTICE | Write the prime factorization of each number below.

4. 40 = _____

5. 105 = _____

6. 66 = _____

7. 43 = _____

8. 128 = _____

9. 222 = _____

10. 364 = _____

11. 475 = _____

PRACTICE | Answer each question below.

12. What is the largest 2-digit prime?

12. _____

13. How many numbers between 200 and 300 have 7 as a prime factor?

13. _____

14. How many factors of 72 are also multiples of 6?

14. _____

15. How many different perimeters are possible for a rectangle with integer side lengths and an area of 126 square units?

15. _____

16. ★ What is the smallest integer that is a factor of $2{,}772 = 2^2 \cdot 3^2 \cdot 7 \cdot 11$, but is *not* a factor of $4{,}200 = 2^3 \cdot 3 \cdot 5^2 \cdot 7$?

16. _____

EXAMPLE | Which of the numbers below is divisible by 28?

$10,540 = 2^2 \cdot 5 \cdot 17 \cdot 31$ $12,936 = 2^3 \cdot 3 \cdot 7^2 \cdot 11$ $14,406 = 2 \cdot 3 \cdot 7^4$

The prime factorization of 28 is $2^2 \cdot 7$. So, a number that is divisible by 28 must have at least two 2's and one 7 in its prime factorization.

$10,540 = 2^2 \cdot 5 \cdot 17 \cdot 31$:
Since there is no 7 in the prime factorization of 10,540, it is not divisible by 28.

$12,936 = 2^3 \cdot 3 \cdot 7^2 \cdot 11$:
Since the prime factorization of 12,936 has at least two 2's and one 7, it is divisible by 28.
$12,936 = 2^3 \cdot 3 \cdot 7^2 \cdot 11 = (2 \cdot 2 \cdot 7) \cdot 2 \cdot 3 \cdot 7 \cdot 11 = (28) \cdot 2 \cdot 3 \cdot 7 \cdot 11$.

$14,406 = 2 \cdot 3 \cdot 7^4$:
Since there is only one 2 in the prime factorization of 14,406, it is not divisible by 28.

So, the only number above that is divisible by 28 is **12,936**.

PRACTICE | Answer each question below.

17. Circle every number below that is divisible by 54.

$675 = 3^3 \cdot 5^2$ $882 = 2 \cdot 3^2 \cdot 7^2$ $1,782 = 2 \cdot 3^4 \cdot 11$ $2,160 = 2^4 \cdot 3^3 \cdot 5$

18. Circle every number below that is divisible by 308.

$3,360 = 2^5 \cdot 3 \cdot 5 \cdot 7$ $4,312 = 2^3 \cdot 7^2 \cdot 11$ $6,468 = 2^2 \cdot 3 \cdot 7^2 \cdot 11$ $9,702 = 2 \cdot 3^2 \cdot 7^2 \cdot 11$

19. Circle every number below that is a factor of $4,095 = 3^2 \cdot 5 \cdot 7 \cdot 13$.

27   91   105   225   315

20. Circle every number below that is a factor of $5,472 = 2^5 \cdot 3^2 \cdot 19$.

32   48   108   119   171

EXAMPLE | The prime factorization of 11,625 is $3 \cdot 5^3 \cdot 31$.
What is $11,625 \div 155$?

The prime factorization of 155 is $5 \cdot 31$.

We can use prime factorization to write 11,625 as the product of 155 and another integer:

$$\begin{aligned} 11,625 &= 3 \cdot 5^3 \cdot 31 \\ &= 3 \cdot 5 \cdot 5 \cdot 5 \cdot 31 \\ &= (5 \cdot 31) \cdot (3 \cdot 5 \cdot 5) \\ &= 155 \cdot 75. \end{aligned}$$

Since $11,625 = 155 \cdot \boxed{75}$, we have $11,625 \div 155 = \boxed{75}$.

PRACTICE | Answer each question below.

21. The prime factorization of 7,425 is $3^3 \cdot 5^2 \cdot 11$. What is $7,425 \div 75$?

21. _____

22. The prime factorization of 49,392 is $2^4 \cdot 3^2 \cdot 7^3$. What is $49,392 \div 196$?

22. _____

23. The prime factorization of 3,780 is $2^2 \cdot 3^3 \cdot 5 \cdot 7$. What number can be multiplied by 135 to get 3,780?

23. _____

24. ★ Ivan divides 504 by its largest odd factor. What is the result?

24. _____

25. ★ The prime factorization of 64,800 is $2^5 \cdot 3^4 \cdot 5^2$. What is the smallest integer quotient Myrtle can get if she divides 64,800 by a power of 6?

25. _____

EXAMPLE | Which of the following are perfect squares?

$$2{,}401 = 7^4 \qquad 21{,}296 = 2^4 \cdot 11^3 \qquad 28{,}900 = 2^2 \cdot 5^2 \cdot 17^2$$

A perfect square is the product of an integer and itself. So, we try to split the prime factorizations above into two identical groups of prime factors.

We can do this for 2,401 and for 28,900:

$$
\begin{aligned}
2{,}401 &= 7^4 \\
&= 7 \cdot 7 \cdot 7 \cdot 7 \\
&= (7 \cdot 7) \cdot (7 \cdot 7) \\
&= 49 \cdot 49 \\
&= 49^2.
\end{aligned}
\qquad\qquad
\begin{aligned}
28{,}900 &= 2^2 \cdot 5^2 \cdot 17^2 \\
&= 2 \cdot 2 \cdot 5 \cdot 5 \cdot 17 \cdot 17 \\
&= (2 \cdot 5 \cdot 17) \cdot (2 \cdot 5 \cdot 17) \\
&= (170) \cdot (170) \\
&= 170^2.
\end{aligned}
$$

Review perfect squares in Chapter 5 of Beast Academy 3B.

However, $\begin{aligned}21{,}296 &= 2^4 \cdot 11^3 \\ &= 2 \cdot 2 \cdot 2 \cdot 2 \cdot 11 \cdot 11 \cdot 11 \\ &= (2 \cdot 2 \cdot 11) \cdot (2 \cdot 2 \cdot 11) \cdot 11.\end{aligned}$

There is no way to group the last 11 so that we have two identical groups of factors.

So, only **2,401** and **28,900** are perfect squares.

PRACTICE | Answer each question below.

26. Write the prime factorization of each perfect square below.

81 = _____ 121 = _____

1,600 = _____ 3,600 = _____

27. Write each prime factorization below as a perfect square.

Ex: $2^4 \cdot 5^2 = \underline{20^2}$ $\qquad 3^2 \cdot 11^2 = $ _____ $\qquad 2^8 = $ _____ $\qquad 2^4 \cdot 3^2 \cdot 7^2 = $ _____

28. Circle every number below that is a perfect square.

$7{,}776 = 2^5 \cdot 3^5 \qquad 3{,}136 = 2^6 \cdot 7^2 \qquad 81{,}796 = 2^2 \cdot 11^2 \cdot 13^2 \qquad 444{,}771 = 3^4 \cdot 17^2 \cdot 19$

PRACTICE | Answer each question below.

29. Circle each number below that is a perfect square when x and y are different prime numbers.

$$x^5 \cdot y^5 \qquad x^{81} \qquad x^2 \cdot y^3 \qquad y^{12} \qquad x \cdot y \qquad x^4 \cdot y^{10}$$

30. Grogg says that if a number is a perfect square, then its prime factorization includes only even exponents. Lizzie says that if the prime factorization of a number includes only even exponents, then the number is a perfect square. Who is correct: Grogg, Lizzie, or both? Explain.

31. Is 9^3 a perfect square? If so, write 9^3 as a perfect square. If not, explain why not.

32. What is the smallest positive integer n for which $180n$ is a perfect square?

32. _____

33. The prime factorization of 6,174 is $2 \cdot 3^2 \cdot 7^3$. What is the smallest positive integer that can be multiplied by 6,174 to make a perfect square?

33. _____

34. What is the largest perfect square factor of $23{,}520 = 2^5 \cdot 3 \cdot 5 \cdot 7^2$?

34. _____

The **Greatest Common Factor (GCF)** of two or more positive integers is the largest number that is a factor of each integer.

EXAMPLE | What is the GCF of 180 and 234?

We list all of the factors of both numbers.

We see that 18 is the largest number that appears in both lists. So, **18** is the GCF of 180 and 234.

180		234
1 · 180	6 · 30	1 · 234
2 · 90	9 · 20	2 · 117
3 · 60	10 ·⑱	3 · 78
4 · 45	12 · 15	6 · 39
5 · 36		9 · 26
		13 ·⑱

— *or* —

We use the prime factorizations of 180 and 234:

$$180 = 2^2 \cdot 3^2 \cdot 5 \qquad 234 = 2 \cdot 3^2 \cdot 13$$

We write the prime factorizations with their common prime factors grouped:

$$180 = 2^2 \cdot 3^2 \cdot 5 = (2 \cdot 3^2) \cdot 2 \cdot 5$$
$$234 = 2 \cdot 3^2 \cdot 13 = (2 \cdot 3^2) \cdot 13$$

The GCF of 180 and 234 is the product of all the prime factors they share. So, the GCF is $2 \cdot 3^2 = $ **18**.

$$180 = (2 \cdot 3^2) \cdot 2 \cdot 5 = (18) \cdot 2 \cdot 5$$
$$234 = (2 \cdot 3^2) \cdot 13 = (18) \cdot 13$$

We sometimes use the notation GCF(a, b) to mean "the GCF of a and b."

PRACTICE | Compute the GCF for each pair of numbers below.

35. GCF(12, 15) = _____

36. GCF(40, 56) = _____

37. GCF(98, 168) = _____

38. GCF(63, 100) = _____

39. GCF(245, 315) = _____

40. GCF(99, 495) = _____

PRACTICE | Answer each question below.

41. What is the largest integer that is a factor of 150, 210, and 525?

41. _____

42. If n is a positive integer, which value below represents the GCF of n and $3n$?

$$1 \qquad n \qquad 3n \qquad \frac{n}{2} \qquad \frac{n}{3}$$

43. If k is a positive integer, which value below that represents the GCF of $12k$ and $18k$?

$$6 \qquad k \qquad 2k \qquad 3k \qquad 6k$$

44. ★ If r is an integer, and the GCF of $6r$ and $8r$ is 44, then what is r?

44. $r =$ _____

45. ★ The GCF of two numbers, a and b, is 180. What is the **second-largest** factor that a and b have in common?

45. _____

46. ★ The greatest common factor of 1,008 and 1,620 is 36. List **every** common factor of 1,008 and 1,620.

46. _____

The GCF of any two numbers is the same as the GCF of either number and their difference. For example, the GCF of 70 and 119 is the same as the GCF of 70 and their difference, 49. We can use this to simplify tough GCF problems.

EXAMPLE | Find the GCF of 77 and 132.

The GCF of 77 and 132 is the same as the GCF of 77 and $132 - 77 = 55$.

Then, since $77 = 7 \cdot 11$ and $55 = 5 \cdot 11$, the GCF of 77 and 55 is 11.

So, the GCF of 77 and 132 is **11**.

Why does this work?

Since 11 is a common factor of 77 and 132, we can skip-count to 132 by 11's, landing on 77 along the way.

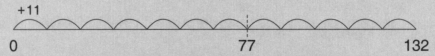

So, we can use 11 to skip-count from 0 to 77.

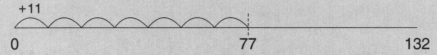

We can also use 11 to skip-count the remaining $132 - 77 = 55$ units from 77 to 132.

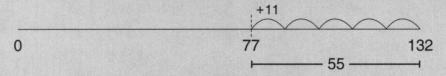

So, since 11 is a factor of 77 and 132, it is also a factor of 55. Therefore, 11 is a common factor of 77 and 55. Similarly, any common factor of 77 and 132 is also a factor of 55.

Additionally, any common factor of 77 and 55 can be used to skip-count from 0 to 132, stopping at 77, as shown below.

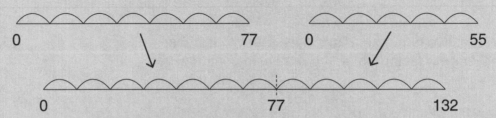

So, any common factor of 77 and 55 is also a factor of 132.

We've shown that all the common factors of 77 and 132 are factors of 55, and that all the common factors of 77 and 55 are factors of 132. So, the common factors of 77 and 132 are the same as the common factors of 77 and 55.

Therefore, the GCF of 77 and 132 is the same as the GCF of 77 and 55:

$$\text{GCF}(77, 132) = \text{GCF}(77, 132 - 77) = \text{GCF}(77, 55).$$

PRACTICE | Use the strategy described on the previous page to compute the GCF of each pair of numbers below.

47. GCF(68, 70) = _____

48. GCF(360, 372) = _____

49. GCF(70, 91) = _____

50. GCF(81, 126) = _____

51. GCF(114, 38) = _____

52. GCF(191, 210) = _____

On the previous page, we showed that GCF(a, b) = GCF(a, a−b) for integers a and b. We can use the same reasoning to show that GCF(a, b) = GCF(a, a+b).

For example, GCF(15, 10) = GCF(15, 15+10) = GCF(15, 25).

PRACTICE | Answer each question below.

53. What is the GCF of 57 and 5,643?

53. _____

54. What is the GCF of 222 and 778?

54. _____

55. ★ 🖉 Is GCF(a, b) **always** equal to GCF($a+b, a-b$)? If yes, explain why. If no, provide an example of two integers a and b where GCF(a, b) does not equal GCF($a+b, a-b$).

The Euclidean Algorithm

EXAMPLE | What is the GCF of 196 and 336?

Using the strategy we learned on the previous two pages, we know that
GCF(196, 336) = GCF(196, 336 − 196) = GCF(196, 140).

However, the GCF of 196 and 140 is not obvious. So, we repeat the same step of replacing the larger number with the difference between the two numbers. We can keep doing this until the GCF is easy to compute!

Replacing 196 with 196 − 140 = 56, we have GCF(196, 140) = GCF(56, 140).
Replacing 140 with 140 − 56 = 84, we have GCF(56, 140) = GCF(56, 84).
Replacing 84 with 84 − 56 = 28, we have GCF(56, 84) = GCF(56, 28).
Replacing 56 with 56 − 28 = 28, we have GCF(56, 28) = GCF(28, 28).

So, GCF(196, 336) = GCF(28, 28) = **28**.

We don't have to keep goin' 'til we get to GCF(28, 28).

We can stop subtractin' once the GCF is obvious!

"Yoo-clid-ian al-go-rhythm."

*This process for finding the GCF of two numbers by repeatedly replacing the larger number with the difference between it and the smaller number is called the **Euclidean Algorithm**.*

It is named after the great math beast Euclid, who first described it thousands of years ago. This is the method most computers use to compute GCFs today!

PRACTICE | Use the Euclidean Algorithm to compute the GCF of each pair of numbers below.

56. GCF(48, 120) = _____

57. GCF(56, 98) = _____

58. GCF(216, 126) = _____

59. GCF(245, 385) = _____

60. Find the GCF of each pair of numbers using the Euclidean Algorithm.

a. GCF(30, 186) = _____   **b.** GCF(360, 51) = _____   **c.** GCF(42, 429) = _____

61. Alex finds the GCF of 273 and 27 with the work shown on the right. Describe a shortcut that would allow Alex to skip Steps 1 through 9.

1. $GCF(273, 27) = GCF(246, 27)$
2. $= GCF(219, 27)$
3. $= GCF(192, 27)$
4. $= GCF(165, 27)$
5. $= GCF(138, 27)$
6. $= GCF(111, 27)$
7. $= GCF(84, 27)$
8. $= GCF(57, 27)$
9. $= GCF(30, 27)$
10. $= GCF(3, 27)$
11. $= \boxed{3}$.

PRACTICE | Use your observations from above to compute the GCF of each of the following pairs of numbers.

62. GCF(42, 434) = _____   **63.** GCF(120, 2424) = _____

64. GCF(90, 912) = _____   **65.** GCF(211, 1085) = _____

The *Least Common Multiple (LCM)* of two or more positive integers is the smallest number that is a multiple of each integer.

EXAMPLE | What is the LCM of 90 and 120?

We list the first few positive multiples of both numbers.

Multiples of 90: 90, 180, 270, (360) 450, ...

Multiples of 120: 120, 240, (360) 480, ...

360 is the smallest number that is in both lists. So, **360** is the LCM of 90 and 120.

— *or* —

We use the prime factorizations of 90 and 120.

$90 = 2 \cdot 3^2 \cdot 5$, so every multiple of 90 must have at least one 2, two 3's, and one 5 in its prime factorization.

$120 = 2^3 \cdot 3 \cdot 5$, so every multiple of 120 must have at least three 2's, one 3, and one 5 in its prime factorization.

So, a multiple of **both** numbers must have at least three 2's, two 3's, and one 5 in its prime factorization. The least common multiple is the number with only these prime factors: $2^3 \cdot 3^2 \cdot 5 = $ **360**.

We have
$360 = 2^3 \cdot 3^2 \cdot 5 = (2 \cdot 3^2 \cdot 5) \cdot 2^2 = 90 \cdot 4$, and
$360 = 2^3 \cdot 3^2 \cdot 5 = (2^3 \cdot 3 \cdot 5) \cdot 3 = 120 \cdot 3$.

We sometimes use the notation LCM(a, b) to mean "the LCM of a and b."

PRACTICE | Compute the LCM for each pair of numbers below.

66. LCM(12, 14) = _____

67. LCM(32, 40) = _____

68. LCM(20, 21) = _____

69. LCM(140, 160) = _____

70. LCM(28, 336) = _____

71. LCM(33, 70) = _____

PRACTICE | Answer each question below.

72. What is the smallest positive integer that is a multiple of 18, 24, and 30?　　72. _____

73. What is the smallest number that is a multiple of each of the first 10 positive integers?　　73. _____

74. What is the **second-smallest** positive integer that is a multiple of 28 and 35?　　74. _____

75. ★ How many positive integers less than 500 are divisible by 4, 5, and 6?　　75. _____

76. If n is a positive integer, which value below represents the LCM of n and $4n$?

　　　　1　　　　n　　　　$4n$　　　　$5n$　　　　$4n^2$

77. If k is a positive integer, which value below represents the LCM of $6k$ and $9k$?

　　　　18　　　　$3k$　　　　$9k$　　　　$18k$　　　　$54k$

78. ★ If p is an integer, and the LCM of $12p$ and $18p$ is 720, then what is p?　　78. $p =$ _____

We can use a **Venn diagram** to organize the prime factors of 2 or 3 numbers.

EXAMPLE | Fill in the Venn diagram on the right with the prime factors of 675 and 1,050.

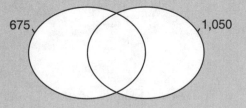

We begin with the prime factorizations of 675 and 1,050.

$$675 = 3^3 \cdot 5^2 \qquad\qquad 1{,}050 = 2 \cdot 3 \cdot 5^2 \cdot 7$$

We fill in the Venn diagram so that shared primes go in the overlapping section. 675 and 1,050 share one 3 and two 5's in their prime factorizations. So, we fill the overlapping section as shown:

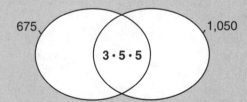

Review Venn Diagrams in Chapter 4 of Beast Academy 4B.

The remaining primes are placed so that the product of the primes within the left oval is 675, and the product of the primes within the right oval is 1,050:

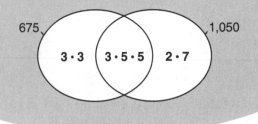

PRACTICE | Answer each question below.

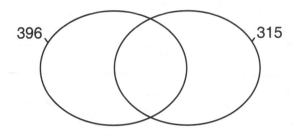

79. Fill in the Venn diagram on the right with the prime factors of 396 and 315.

80. What is the GCF of 396 and 315?

80. _____

81. How can the Venn diagram above be used to find the GCF of 396 and 315?

PRACTICE | Answer each question below.

82. Fill in the Venn diagram on the right with the prime factors of 132 and 231.

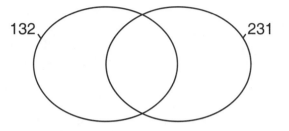

83. What is the LCM of 132 and 231?

83. _____

84. How can the Venn diagram above be used to find the LCM of 132 and 231?

85. Fill in the Venn diagram on the right with the prime factors of 72 and 112.

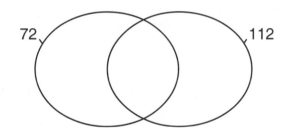

86. Use the Venn diagram above to complete the following statement:

The GCF of 72 and 112 is _____ and the LCM of 72 and 112 is _____.

87. What is the product of the GCF and the LCM you found in the previous problem?

87. _____

88. Compute 72 · 112. Compare this to your previous answer. Why is the product of a pair of numbers always equal to the product of their GCF and their LCM?

89. Two numbers have a product of 490 and a GCF of 7. What is the LCM of the two numbers?

89. _____

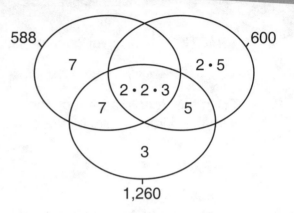

PRACTICE | Use the Venn diagram on the right to answer the questions that follow.

90. What is the prime factorization of 600?

90. _____

91. What is the GCF of 588, 600, and 1,260?

91. _____

92. What is the LCM of 588, 600, and 1,260?

92. _____

93. What is the GCF of 588 and 1,260?

93. _____

94. What is the LCM of 600 and 1,260?

94. _____

95. Complete the Venn diagram below for the integers 120, 175, and 210.

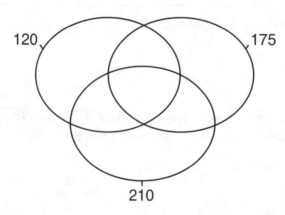

96. Complete the statement below:

The GCF of 120, 175, and 210 is _____ and the LCM of 120, 175, and 210 is _____.

PRACTICE | Answer each question below.

97. Adult worlumps croak every 15 seconds. Baby worlumps croak every 9 seconds. If an adult and a baby worlump just croaked at the same time, after how many seconds will the two croak together again?

97. _____

98. Beast burger patties come in packages of 8. Beast burger buns come in packages of 6. Kruggle buys an equal number of patties and buns. What is the smallest number of packages of buns he could have bought?

98. _____

99. Ren has a sack of snargles worth a total of $126. Ben has a sack of snargles worth a total of $182. If all snargles are worth the same whole number of dollars, what is the largest possible value of a snargle?

99. _____

100. Gardina has 40 apples, 28 oranges, and 36 bananas. What is the largest number of identical fruit baskets Gardina can fill using all of her fruit?

100. _____

PRACTICE | Answer each question below.

101. Byath Leet runs every 6 days and swims every 10 days. He calls it a "super-day" when he runs and swims on the same day. If Byath Leet's 10th super-day is today, how many days will it be until he has his 20th super-day?

101. _____

102. At Acrobeast Gymnasium, one ninth of the gymnasts have a tail, one twelfth of the gymnasts have fur, and one fifteenth of the gymnasts have webbed feet. What is the smallest number of gymnasts that could attend Acrobeast Gymnasium?

★

102. _____

103. The number of stairs between consecutive floors at the Beast Island Hotel is always the same. There are 207 stairs between the bottom floor and the pool floor, and there are 368 stairs between the bottom floor and the top floor. If the bottom floor is floor 1, then what is the number assigned to the pool floor?

★

103. _____

104. Niki plays a game with a standard die. Every time she rolls a prime number she gets 45 points. Every time she **does not** roll a prime number she loses 12 points. At the end of a game, Niki has 0 points. What is the smallest number of rolls she could have made?

★

104. _____

In a **GCF-LCM Web**, we fill in circles as shown in the diagram to the right. We fill the circle **above** the circles labeled a and b with the **LCM** of a and b. We fill the circle **below** the circles labeled a and b with the **GCF** of a and b.

No two circles in a GCF-LCM Web have the same value.

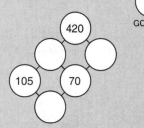

EXAMPLE | Fill each blank to complete the GCF-LCM Web to the right.

We begin by filling in the circles connected to both 105 and 70.

Since $105 = 3 \cdot 5 \cdot 7$, and $70 = 2 \cdot 5 \cdot 7$, we have GCF(105, 70) $= 5 \cdot 7 = $ **35**, and LCM(105, 70) $= 2 \cdot 3 \cdot 5 \cdot 7 = $ **210**.

We fill these circles as shown to the right.

We know that 420 is the LCM of 210 and the missing number. Since 420 has 2^2 in its prime factorization, but $210 = 2 \cdot 3 \cdot 5 \cdot 7$ does not, the missing number must include 2^2 in its prime factorization.

We also know the missing number is a multiple of $70 = 2 \cdot 5 \cdot 7$. So, its prime factorization must include $2 \cdot 5 \cdot 7$.

We see that $2^2 \cdot 5 \cdot 7 = 140$ includes both 2^2 and $2 \cdot 5 \cdot 7$. Including any additional prime factors would cause the LCM to be greater than 420.

So, **140** is the only number that can fill the remaining blank.

PRACTICE | Fill each blank circle to complete the GCF-LCM Webs below.

105.

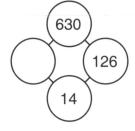

106.

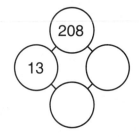

107.

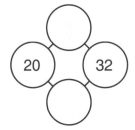

108.

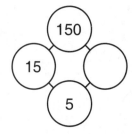

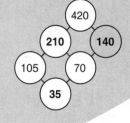

FACTORS & MULTIPLES

Remember, no two circles in a web can have the same value!

PRACTICE | Fill each blank circle to complete the GCF-LCM Webs below.

109.

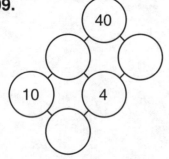

110.

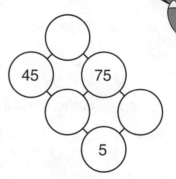

111.

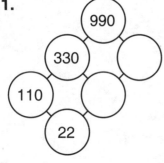

112.

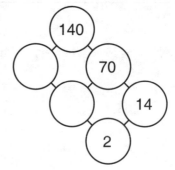

113.

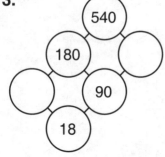

114.
★

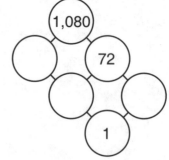

PRACTICE | Fill each blank circle to complete the GCF-LCM Webs below.

115.
★

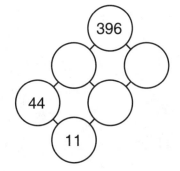

116.
★

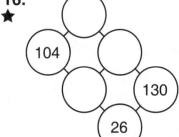

117.

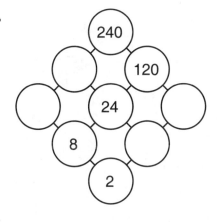

118.

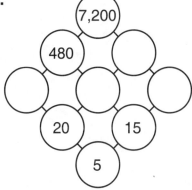

119.
★

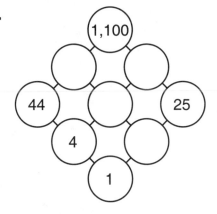

120.
★

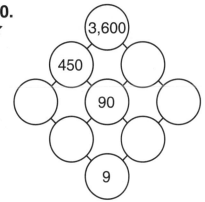

EXAMPLE | What is the power of 3 in the prime factorization of (9!)?

The expression 9! means $9 \cdot 8 \cdot 7 \cdot 6 \cdot 5 \cdot 4 \cdot 3 \cdot 2 \cdot 1$. Writing each composite factor as a product of primes, then grouping primes, we have

$$9! = 9 \cdot 8 \cdot 7 \cdot 6 \cdot 5 \cdot 4 \cdot 3 \cdot 2 \cdot 1$$
$$= (3 \cdot 3) \cdot (2 \cdot 2 \cdot 2) \cdot 7 \cdot (2 \cdot 3) \cdot 5 \cdot (2 \cdot 2) \cdot 3 \cdot 2$$
$$= 2^7 \cdot 3^4 \cdot 5 \cdot 7.$$

So, the power of 3 in the prime factorization of 9! is 3^4.

— *or* —

There are three multiples of 3 from 1 to 9, each of which contributes a 3 to the prime factorization of (9!).

$$9! = ⑨ \cdot 8 \cdot 7 \cdot ⑥ \cdot 5 \cdot 4 \cdot ③ \cdot 2 \cdot 1$$

However, $9 = 3^2$ has *two* 3's in its prime factorization, so it contributes a factor of 3 we have not yet counted. Therefore, there are $3 + 1 = 4$ threes in the prime factorization of 9!. So, the power of 3 in 9! is 3^4.

PRACTICE | Answer each question below.

121. What is the power of 2 in the prime factorization of (7!)?

121. _____

122. What is the power of 3 in the prime factorization of (18!)?

122. _____

123. Find the power of 2 in the prime factorization of each number below:

4!: _____   8!: _____   16!: _____   32!: _____

124. ★ What is the largest power of 6 that is a factor of (12!)?

124. _____

PRACTICE | Answer each question below.

125. What is the largest power of 2 that is a factor of (25!)?

125. _____

126. What is the largest power of 5 that is a factor of (25!)?

126. _____

127. What is the largest power of 10 that is a factor of (25!)?

127. _____

128. ★ A number with trailing zeros has zeros as its rightmost digits. For example, 90,000 has four trailing zeros, and 100,400 has two trailing zeros. How many trailing zeros are there in (25!)?

128. _____

129. ★ How many trailing zeros are there in (100!)?

129. _____

We call two numbers *relatively prime* if their GCF is 1.

EXAMPLE | Which of the following numbers are relatively prime to 12?

77   16   27   35   75

Two numbers that are relatively prime have a GCF of 1. Therefore, the two numbers do not share any prime factors.

The prime factorization of 12 is $2^2 \cdot 3$.

Since each of $16 = 2^4$, $27 = 3^3$, and $75 = 3 \cdot 5^2$ has a 2 or 3 in its prime factorization, none of these numbers are relatively prime to 12.

$77 = 7 \cdot 11$ and $35 = 5 \cdot 7$ share no primes with 12.
So, **77** and **35** are relatively prime to 12.

PRACTICE | Answer each question below.

130. Circle the numbers below that are relatively prime to 7.

7   9   15   28   75

131. Circle the numbers below that are relatively prime to 45.

8   18   21   44   80

132. In the diagram below, connect each pair of relatively prime integers with a line segment.

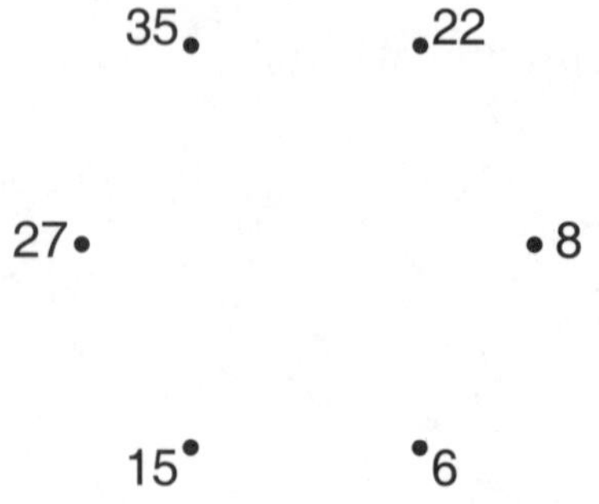

35.   .22

27•   •8

15•   •6

PRACTICE | Answer each question below.

133. List every positive integer less than 20 that is relatively prime to 20.

133. _____

134. What is the smallest number greater than 1 that is relatively prime to 210?

134. _____

135. How many integers between 50 and 70 are relatively prime to both 50 and 70?

135. _____

136. Ralph says, "If a and b are relatively prime, then $a+1$ and $b+1$ are also relatively prime." Is Ralph correct? If yes, explain why. If no, provide an example of two numbers a and b where a and b are relatively prime but $a+1$ and $b+1$ are not relatively prime.

137. Cammie says, "The LCM of two relatively prime numbers is always equal to the product of the two numbers." Is Cammie correct? If yes, explain why. If no, give an example of two relatively prime numbers whose product is not equal to their LCM.

PRACTICE | Answer each question below.

138. Compute the GCF of each pair of numbers below:

    a. GCF(9, 10) = _____   **b.** GCF(65, 66) = _____   **c.** GCF(680, 681) = _____

139. If n is a positive integer, are n and $n+1$ always relatively prime? If yes, explain why. If no, provide an example of a positive integer n where n and $n+1$ are not relatively prime.

140. The sum of relatively prime integers a and b is 1,000. Both numbers are greater than 1, and a is greater than b.

    a. What is the smallest possible value of $a-b$?   **a.** _____

    b. What is the largest possible value of $a-b$?   **b.** _____

141. Grogg writes down three consecutive positive integers. How many of these integers must be relatively prime to 3?   **141.** _____

PRACTICE | Answer each question below.

142. What is the largest power of 4 that is a
factor of 2,880?

142. _____

143. The number 12,222 is divisible by 6, 9, and 21.
★ What is the next-largest integer that is also divisible
by 6, 9, and 21?

143. _____

144. List the two different pairs of integers that
★ have a GCF of 6 and an LCM of 72.

144. _____ & _____

145. The prime factorization of a number is $2^{33} \cdot 5^{28}$. What is the leftmost
★ digit of the number?

145. _____

PRACTICE | Answer each question below.

146. The number $n!$ ends in exactly 7 zeros. What is the smallest possible value of n?

146. _____

147. The product of 10! and a positive integer k is a perfect square. What is the smallest possible value of k?

147. _____

148. Lizzie computes the LCM of 168 and 980. She then includes a third number and computes a *different* LCM for all three numbers. What is the smallest possible value of the third number?

148. _____

149. The LCM of x and y is 120. Find the smallest possible sum $x+y$.

149. _____

PRACTICE | Answer each question below.

150. Alex and Grogg each cut out rectangles whose side lengths are a
★ whole number of inches. Alex's rectangle has area 48 sq in, and
Grogg's rectangle has area 72 sq in. Alex and Grogg connect their
rectangles to create a larger rectangle. What is the smallest possible
perimeter of the new rectangle?

150. _____

151. Circle three numbers below so that the GCF of *any two* of the circled
★ numbers is greater than 1, but the GCF of *all three* circled numbers is 1.

12   14   15   20   30

152. One bell rings every 4 hours, a second bell rings every 6 hours, and
★ a third bell rings every 10 hours. If all three bells just rang together,
★ how many times in the next 100 hours will *exactly two* bells ring
together (including the 100th hour)?

152. _____

153. Captain Kraken has a satchel of gold and silver coins. The gold coins
★ are worth $50 each, while the silver coins are worth $12 each. The
★ average value of a coin in the satchel is $30. What is the smallest
total number of coins that could be in Kraken's satchel?

153. _____

CHAPTER 6
Fractions

Use this Practice book with
Guide 4C from BeastAcademy.com.

Recommended Sequence:

Book	Pages:
Guide:	76-83
Practice:	65-78
Guide:	84-95
Practice:	79-88
Guide:	96-109
Practice:	89-97

You may also read the entire chapter
in the Guide before beginning the
Practice chapter.

EXAMPLE | Compute $\frac{3}{7} + \frac{2}{7}$.

Adding 3 sevenths to 2 sevenths, we get a total of $3+2 = 5$ sevenths.

So, $\frac{3}{7} + \frac{2}{7} = \frac{5}{7}$.

EXAMPLE | Compute $\frac{7}{9} - \frac{5}{9}$.

Subtracting 5 ninths from 7 ninths leaves $7-5 = 2$ ninths.

So, $\frac{7}{9} - \frac{5}{9} = \frac{2}{9}$.

Every answer in this chapter should be in **simplest form**.

Review adding and subtracting fractions in Chapter 8 of Beast Academy 4C!

PRACTICE | Compute each sum or difference below.

1. $\frac{8}{17} + \frac{7}{17} =$

2. $\frac{6}{13} - \frac{4}{13} =$

3. $\frac{3}{10} + \frac{9}{10} =$

4. $\frac{3}{4} - \frac{1}{4} =$

5. $4\frac{5}{9} + 3\frac{2}{9} =$

6. $7\frac{4}{5} + 6\frac{3}{5} =$

7. $11\frac{7}{8} - 2\frac{5}{8} =$

8. $16 - 4\frac{2}{7} =$

FRACTIONS
Review

EXAMPLE | What is $\frac{7}{11}$ of 13?

To find $\frac{7}{11}$ of 13, we multiply $\frac{7}{11} \cdot 13$.

$$\frac{7}{11} \cdot 13 = \frac{7 \cdot 13}{11} = \frac{91}{11}.$$

So, $\frac{7}{11}$ of 13 is $\frac{91}{11} = 8\frac{3}{11}$.

Review multiplying fractions and whole numbers in Chapter 10 of Beast Academy 4D!

Sometimes, rearranging an expression can make a computation easier.

EXAMPLE | Compute $21 \cdot \frac{5}{7}$.

We notice that 21 is a multiple of 7, so we have

$$21 \cdot \frac{5}{7} = \frac{21 \cdot 5}{7}$$
$$= \frac{21}{7} \cdot 5$$
$$= 3 \cdot 5$$
$$= \mathbf{15}.$$

PRACTICE | Compute each product below.

9. What is $\frac{2}{9}$ of 54?

10. Compute $\frac{5}{6} \cdot 11$.

9. _____

10. _____

11. What is $\frac{3}{8}$ of 7?

12. Compute $\frac{5}{28} \cdot 14$.

11. _____

12. _____

13. $\frac{7}{12} \cdot 36 =$ _____

14. $26 \cdot \frac{6}{13} =$ _____

15. $9 \cdot \frac{8}{15} =$ _____

16. $3\frac{1}{2} \cdot 7 =$ _____

To add or subtract fractions with different denominators, we rewrite the fractions so that they have the same denominator.

EXAMPLE | Compute $\frac{2}{5} + \frac{3}{10}$.

We begin by converting $\frac{2}{5}$ into tenths: $\frac{2}{5} = \frac{4}{10}$.

Then, we add: $\frac{2}{5} + \frac{3}{10} = \frac{4}{10} + \frac{3}{10} = \frac{\mathbf{7}}{\mathbf{10}}$.

PRACTICE | Compute each sum or difference below.

17. $\frac{3}{5} + \frac{2}{15} =$

18. $\frac{5}{8} - \frac{5}{16} =$

19. $\frac{5}{12} + \frac{1}{4} =$

20. $\frac{5}{9} - \frac{1}{3} =$

21. $\frac{17}{18} - \frac{5}{6} =$

22. $\frac{1}{2} + \frac{3}{14} =$

23. $4\frac{2}{3} - 2\frac{5}{21} =$

24. $12\frac{61}{63} + 5\frac{2}{7} =$

FRACTIONS
Different Denominators

Sometimes, we need to convert **both** fractions in a sum or
difference so that they have the same denominator.

EXAMPLE | Compute $\frac{5}{9} - \frac{4}{15}$.

Since $9 \cdot 15 = 135$ is a common multiple of both denominators, we
can begin by converting both $\frac{5}{9}$ and $\frac{4}{15}$ into 135ths:

$$\frac{5}{9} \overset{\cdot 15}{\underset{\cdot 15}{=}} \frac{75}{135} \qquad \frac{4}{15} \overset{\cdot 9}{\underset{\cdot 9}{=}} \frac{36}{135}$$

Then, we subtract: $\frac{5}{9} - \frac{4}{15} = \frac{75}{135} - \frac{36}{135} = \frac{39}{135} = \frac{\mathbf{13}}{\mathbf{45}}$.

— *or* —

Since $9 = 3^2$ and $15 = 3 \cdot 5$, the LCM of 9 and 15 is $3^2 \cdot 5 = 45$.
So, we convert both $\frac{5}{9}$ and $\frac{4}{15}$ into 45ths:

$$\frac{5}{9} \overset{\cdot 5}{\underset{\cdot 5}{=}} \frac{25}{45} \qquad \frac{4}{15} \overset{\cdot 3}{\underset{\cdot 3}{=}} \frac{12}{45}$$

Then, we subtract: $\frac{5}{9} - \frac{4}{15} = \frac{25}{45} - \frac{12}{45} = \frac{\mathbf{13}}{\mathbf{45}}$.

It doesn't matter which common multiple we choose for the denominator...
...but the LCM helps us keep the numbers as small as possible!

PRACTICE | Compute each sum or difference below.

25. $\frac{1}{3} + \frac{1}{4} =$

26. $\frac{3}{4} - \frac{5}{14} =$

27. $\frac{2}{5} - \frac{1}{4} =$

28. $\frac{1}{6} + \frac{7}{10} =$

29. $\frac{1}{6} + \frac{8}{9} =$

30. $3\frac{4}{9} - 1\frac{3}{10} =$

PRACTICE | Answer each question below.

31. Frank ran $2\frac{5}{6}$ miles and Dave ran $3\frac{7}{8}$ miles. How many miles farther did Dave run than Frank?

31. _____

32. Inchie the Worm climbs $4\frac{1}{3}$ inches up a rope, slips down $1\frac{1}{2}$ inches, and then climbs up $3\frac{3}{8}$ inches more. How many inches is Inchie from his starting place on the rope?

32. _____

33. What is the perimeter in feet of a rectangle with height $3\frac{1}{4}$ feet and width $3\frac{1}{3}$ feet?

33. _____

34. All measurements below are given in meters. The perimeter of the pentagon is $9\frac{9}{20}$ meters. What is the length of the unlabeled side?

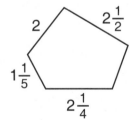

34. _____

35. Tina has a bag of solid-colored marbles. Two fifths of her marbles are red, one third of her marbles are blue, and the rest are yellow. What fraction of Tina's marbles are yellow?

35. _____

36. Compute $\frac{1}{3} - \frac{1}{5} + \frac{1}{5} - \frac{1}{7} + \frac{1}{7} - \frac{1}{9} + \frac{1}{9} - \frac{1}{11} + \frac{1}{11} - \frac{1}{13} + \frac{1}{13} - \frac{1}{15}$.

36. _____

PRACTICE | Answer each question below.

37. Below are the number of centimeters Rosa grew each month.

Jan	Feb	March	April	May	June
$1\frac{1}{3}$	$\frac{5}{6}$	$\frac{1}{2}$	$2\frac{1}{4}$	$\frac{1}{3}$	$1\frac{3}{4}$

a. How many centimeters taller was Rosa at the end of May than she was at the beginning of March?

a. _____

b. Rosa was $157\frac{1}{3}$ cm tall at the end of June. How many centimeters tall was she at the end of April?

b. _____

c. ★ What was Rosa's average monthly growth over these six months, in centimeters?

c. _____

38. ★ A hexatoad and octopug together weigh $10\frac{1}{6}$ pounds. The octopug weighs $1\frac{1}{2}$ pounds more than the hexatoad. How many pounds does the hexatoad weigh?

38. _____

39. ★ Compute $2\left(1-\frac{1}{2}\right)+3\left(1-\frac{2}{3}\right)+\cdots+99\left(1-\frac{98}{99}\right)+100\left(1-\frac{99}{100}\right)$.

39. _____

40. ★ ★ Suppose that A and B are positive integers with A<B and $\frac{1}{A}+\frac{1}{B}=\frac{1}{3}$. What are A and B?

40. A = _____

B = _____

A sum of **distinct** unit fractions is called an **Egyptian fraction**. Ancient Egyptians used unit fractions to represent all other fractions. For example, ancient Egyptians might have written $\frac{2}{3}$ as $\frac{1}{2} + \frac{1}{6}$.

We can write $\frac{7}{12}$ as an Egyptian fraction as shown:

$$\frac{7}{12} = \frac{1}{3} + \frac{1}{6} + \frac{1}{12}.$$

We could also write $\frac{7}{12} = \frac{1}{3} + \frac{1}{4}$ or $\frac{7}{12} = \frac{1}{2} + \frac{1}{12}$.

When we write Egyptian fractions, we order the unit fractions in the sum from greatest to least.

PRACTICE | Write each Egyptian fraction below as a single fraction.

41. $\frac{1}{4} + \frac{1}{8} =$

42. $\frac{1}{5} + \frac{1}{20} =$

43. $\frac{1}{3} + \frac{1}{4} + \frac{1}{12} =$

44. $\frac{1}{2} + \frac{1}{5} + \frac{1}{10} =$

PRACTICE | Fill in the missing denominator in each equation below to show how each fraction can be written as an Egyptian fraction.

45. $\frac{5}{12} = \frac{1}{3} + \frac{1}{\underline{}}$

46. $\frac{3}{11} = \frac{1}{4} + \frac{1}{\underline{}}$

47. $\frac{7}{9} = \frac{1}{2} + \frac{1}{4} + \frac{1}{\underline{}}$

48. $\frac{5}{11} = \frac{1}{3} + \frac{1}{\underline{}} + \frac{1}{99}$

Egyptian Fractions

EXAMPLE | Write $\frac{3}{7}$ as an Egyptian fraction.

We begin by finding the largest unit fraction that can be part of the Egyptian fraction.

$\frac{1}{3}$ is the largest unit fraction that is less than $\frac{3}{7}$. Subtracting $\frac{1}{3}$ from $\frac{3}{7}$ gives us

$\frac{3}{7} - \frac{1}{3} = \frac{9}{21} - \frac{7}{21} = \frac{2}{21}$. So, we have

$$\frac{3}{7} = \frac{1}{3} + \frac{2}{21}.$$

Next, we write $\frac{2}{21}$ as the sum of distinct unit fractions.

Since $\frac{2}{22} < \frac{2}{21} < \frac{2}{20}$, we know $\frac{1}{11} < \frac{2}{21} < \frac{1}{10}$. So, $\frac{1}{11}$ is the largest unit fraction that is less

than $\frac{2}{21}$. Subtracting $\frac{1}{11}$ from $\frac{2}{21}$ gives us $\frac{2}{21} - \frac{1}{11} = \frac{22}{231} - \frac{21}{231} = \frac{1}{231}$. So, we have

$$\frac{2}{21} = \frac{1}{11} + \frac{1}{231}.$$

Now, we can write $\frac{3}{7}$ as a sum of distinct unit fractions:

$$\frac{3}{7} = \frac{1}{3} + \frac{2}{21}$$
$$= \frac{1}{3} + \frac{1}{11} + \frac{1}{231}.$$

Any fraction between 0 and 1 can be written as an Egyptian fraction using this method.

PRACTICE | Write each fraction below as an Egyptian fraction using the method shown above.

49. $\frac{3}{4} = \frac{1}{} + \frac{1}{}$

50. $\frac{2}{3} = \frac{1}{} + \frac{1}{}$

51. $\frac{5}{16} = \frac{1}{} + \frac{1}{}$

52. $\frac{4}{5} = \frac{1}{} + \frac{1}{} + \frac{1}{}$

53. In the example above, we wrote $\frac{3}{7}$ as $\frac{1}{3} + \frac{1}{11} + \frac{1}{231}$.
Fill in the blank denominators at the right to find another
way to write $\frac{3}{7}$ as an Egyptian fraction.

$$\frac{3}{7} = \frac{1}{4} + \frac{1}{} + \frac{1}{}$$

PRACTICE | Find the missing denominator that makes each equation below true.

54. $\dfrac{1}{5} = \dfrac{1}{6} + \dfrac{1}{\rule{1cm}{0.4pt}}$

55. $\dfrac{1}{9} = \dfrac{1}{10} + \dfrac{1}{\rule{1cm}{0.4pt}}$

56. $\dfrac{1}{8} = \dfrac{1}{9} + \dfrac{1}{\rule{1cm}{0.4pt}}$

57. $\dfrac{1}{99} = \dfrac{1}{100} + \dfrac{1}{\rule{1cm}{0.4pt}}$

58. Find an expression for the missing denominator below that makes the equation true.

$$\frac{1}{n} = \frac{1}{n+1} + \frac{1}{\rule{1.5cm}{0.4pt}}$$

> Math beasts still study Egyptian fractions.

> For example, math beasts **think** that any fraction with numerator 4 can be written as a sum of three distinct unit fractions...

> ...but nobody knows for sure!

PRACTICE | Use your answers from above to help you write each fraction below as an Egyptian fraction.

59. $\dfrac{2}{5} = \dfrac{1}{\rule{1cm}{0.4pt}} + \dfrac{1}{\rule{1cm}{0.4pt}}$

60. $\dfrac{2}{9} = \dfrac{1}{\rule{1cm}{0.4pt}} + \dfrac{1}{\rule{1cm}{0.4pt}}$

61. $\dfrac{3}{8} = \dfrac{1}{\rule{1cm}{0.4pt}} + \dfrac{1}{\rule{1cm}{0.4pt}}$

62. $\dfrac{10}{99} = \dfrac{1}{\rule{1cm}{0.4pt}} + \dfrac{1}{\rule{1cm}{0.4pt}}$

63. ★ Find **two different** pairs of unit fractions whose sum is $\dfrac{8}{15}$.

$$\frac{8}{15} = \frac{1}{\rule{1cm}{0.4pt}} + \frac{1}{\rule{1cm}{0.4pt}} \qquad \frac{8}{15} = \frac{1}{\rule{1cm}{0.4pt}} + \frac{1}{\rule{1cm}{0.4pt}}$$

In a **Crossout Sum** puzzle, each row and each column has exactly 2 squares that contain positive fractions. The clues outside the table give the sum of the fractions in each row or column.

EXAMPLE | Complete the Crossout Sum puzzle to the right.

The missing number in the middle column is $\frac{7}{8} - \frac{1}{2} = \frac{7}{8} - \frac{4}{8} = \frac{3}{8}$.

The sum of the two fractions in the bottom row is $\frac{1}{4}$, which is less than $\frac{3}{8}$. So, $\frac{3}{8}$ is not in the bottom row.

So, we place $\frac{3}{8}$ as shown in the center square, and we X out the bottom-middle square as a reminder that this square does not contain a number.

The missing number in the middle row is $\frac{9}{8} - \frac{3}{8} = \frac{6}{8} = \frac{3}{4}$.

The sum of the two fractions in the left column is $\frac{1}{2}$, which is less than $\frac{3}{4}$. So, we place $\frac{3}{4}$ as shown in the middle-right square, and cross out the remaining square in this row.

Each row and each column has exactly 2 squares that contain numbers. So, in this 3-by-3 puzzle, each row and each column has exactly one square that contains an X instead of a number. Therefore, we X out the top-right square as shown.

The top-left number is $\frac{3}{5} - \frac{1}{2} = \frac{6}{10} - \frac{5}{10} = \frac{1}{10}$.

The bottom-right number is $\frac{11}{10} - \frac{3}{4} = \frac{22}{20} - \frac{15}{20} = \frac{7}{20}$.

Finally, we use either the row or column sum to compute the remaining entry.

$$\frac{3}{4} - \frac{7}{20} = \frac{15}{20} - \frac{7}{20} = \frac{8}{20} = \frac{2}{5} \quad \textit{or} \quad \frac{1}{2} - \frac{1}{10} = \frac{5}{10} - \frac{1}{10} = \frac{4}{10} = \frac{2}{5}.$$

PRACTICE | Solve each Crossout Sum puzzle below.

64.

	$\frac{3}{8}$	$\frac{7}{12}$
$\frac{7}{24}$		
$\frac{2}{3}$	$\frac{1}{4}$	

65.

	$\frac{5}{8}$	$\frac{3}{4}$
$\frac{4}{5}$	$\frac{1}{4}$	
$\frac{23}{40}$		

PRACTICE | Solve each Crossout Sum puzzle below.

66.

	$\frac{19}{15}$	$\frac{29}{21}$	$\frac{7}{18}$
$\frac{22}{15}$		$\frac{2}{3}$	
$\frac{37}{42}$			$\frac{1}{6}$
$\frac{31}{45}$	$\frac{7}{15}$		

67.

	$\frac{16}{15}$	$\frac{27}{40}$	$\frac{11}{28}$
$\frac{17}{21}$			$\frac{1}{7}$
$\frac{7}{10}$	$\frac{2}{5}$		
$\frac{5}{8}$		$\frac{3}{8}$	

68.

	$\frac{37}{28}$	$\frac{43}{30}$	$\frac{7}{9}$
$\frac{64}{63}$	$\frac{4}{7}$		
$\frac{27}{20}$		$\frac{3}{5}$	
$\frac{7}{6}$			$\frac{1}{3}$

69.

	$\frac{31}{24}$	$\frac{4}{3}$	$\frac{31}{35}$
$\frac{49}{40}$			$\frac{3}{5}$
$\frac{11}{14}$		$\frac{1}{2}$	
$\frac{3}{2}$	$\frac{2}{3}$		

70.

	$\frac{1}{2}$	$\frac{5}{6}$	$\frac{14}{15}$
$\frac{7}{12}$		✕	
$\frac{19}{15}$		$\frac{2}{3}$	
$\frac{5}{12}$			

71.

	$\frac{79}{70}$	$\frac{25}{24}$	$\frac{13}{15}$
$\frac{5}{6}$			
$\frac{9}{10}$			$\frac{1}{5}$
$\frac{73}{56}$	$\frac{3}{7}$		

EXAMPLE Use the given numbers to fill in the blanks to create a true equation. *All fractions must be less than 1 and in simplest form.*

Numbers: 1, 3, 8, 24

$$\frac{13}{\rule{1cm}{0.4pt}} + \frac{\rule{1cm}{0.4pt}}{\rule{1cm}{0.4pt}} = \frac{2}{\rule{1cm}{0.4pt}}$$

First, we look at $\frac{13}{\rule{0.5cm}{0.4pt}}$. Since every fraction is less than 1, only 24 can be the denominator of $\frac{13}{\rule{0.5cm}{0.4pt}}$.

$$\frac{13}{24} + \frac{\rule{1cm}{0.4pt}}{\rule{1cm}{0.4pt}} = \frac{2}{\rule{1cm}{0.4pt}}$$

The remaining numbers are 1, 3, and 8.
Next, we look at $\frac{2}{\rule{0.5cm}{0.4pt}}$. Every fraction is in simplest form and less than 1. Since $\frac{2}{8}$ can be simplified and $\frac{2}{1}$ is less than 1, only 3 can be the denominator of $\frac{2}{\rule{0.5cm}{0.4pt}}$.

$$\frac{13}{24} + \frac{\rule{1cm}{0.4pt}}{\rule{1cm}{0.4pt}} = \frac{2}{3}$$

The remaining numbers are 1 and 8. There is only one way to place these into the empty numerator and denominator to make a fraction that is less than 1.

$$\frac{13}{24} + \frac{1}{8} = \frac{2}{3}$$

Finally, we check that the equation is true:

$$\frac{13}{24} + \frac{1}{8} = \frac{13}{24} + \frac{3}{24} = \frac{16}{24} = \frac{2}{3}. \checkmark$$

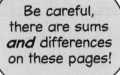

Be careful, there are sums *and* differences on these pages!

PRACTICE For each problem below, use the given numbers to fill in the blanks to create a true equation. *All fractions must be less than 1 and in simplest form.*

72. Numbers: 1, 4, 21

$$\frac{\rule{1cm}{0.4pt}}{6} + \frac{\rule{1cm}{0.4pt}}{\rule{1cm}{0.4pt}} = \frac{5}{14}$$

73. Numbers: 1, 2, 11

$$\frac{\rule{1cm}{0.4pt}}{22} + \frac{5}{\rule{1cm}{0.4pt}} = \frac{1}{\rule{1cm}{0.4pt}}$$

74. Numbers: 1, 4, 5

$$\frac{3}{\rule{1cm}{0.4pt}} - \frac{\rule{1cm}{0.4pt}}{15} = \frac{\rule{1cm}{0.4pt}}{3}$$

75. Numbers: 1, 5, 13

$$\frac{\rule{1cm}{0.4pt}}{6} - \frac{\rule{1cm}{0.4pt}}{18} = \frac{\rule{1cm}{0.4pt}}{9}$$

PRACTICE | For each problem below, use the given numbers to fill in the blanks to create a true equation. ***All fractions must be less than 1 and in simplest form.***

76. Numbers: 3, 4, 17

$$\frac{}{20} - \frac{}{5} = \frac{1}{}$$

77. Numbers: 2, 3, 7, 8, 24

$$\frac{}{} + \frac{}{} = \frac{}{3}$$

78. Numbers: 1, 2, 5, 9

$$\frac{2}{} + \frac{}{} = \frac{}{10}$$

79. Numbers: 5, 6, 9, 15

$$\frac{}{10} - \frac{1}{} = \frac{}{}$$

80. Numbers: 1, 2, 4, 5

$$\frac{}{} + \frac{}{} = \frac{13}{20}$$

81. Numbers: 1, 2, 3, 5

$$\frac{}{} + \frac{}{} = \frac{11}{15}$$

82. ★ Numbers: 2, 3, 7, 21

$$\frac{2}{} - \frac{4}{} = \frac{}{}$$

83. ★ Numbers: 5, 7, 13, 28, 42

$$\frac{}{} + \frac{}{} = \frac{}{12}$$

PRACTICE | Place a "+" or "−" in each blank square to make true statements.

84. $\frac{3}{5}$ ☐ $\frac{1}{3}$ = $\frac{4}{15}$

85. $\frac{1}{9}$ ☐ $\frac{2}{15}$ = $\frac{11}{45}$

86. $\frac{7}{8}$ ☐ $\frac{1}{2}$ ☐ $\frac{1}{8}$ = $\frac{1}{2}$

87. $\frac{5}{12}$ ☐ $\frac{1}{3}$ ☐ $\frac{1}{2}$ = $\frac{1}{4}$

88. $\frac{1}{2}$ ☐ $\frac{1}{4}$ ☐ $\frac{3}{5}$ = $\frac{27}{20}$

89. $\frac{5}{6}$ ☐ $\frac{1}{4}$ ☐ $\frac{2}{5}$ = $\frac{59}{60}$

90. $\frac{13}{18}$ ☐ $\frac{1}{2}$ ☐ $\frac{1}{9}$ = $\frac{1}{3}$

91. $\frac{11}{20}$ ☐ $\frac{19}{24}$ ☐ $\frac{4}{5}$ = $\frac{13}{24}$

EXAMPLE | What number is $\frac{1}{3}$ of $\frac{1}{2}$?

We locate $\frac{1}{2}$ on the number line by splitting the number line between 0 and 1 into two equal pieces.

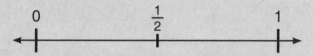

To find $\frac{1}{3}$ of $\frac{1}{2}$, we split each half into 3 equal pieces. Then, there are $2 \cdot 3 = 6$ equal pieces between 0 and 1. Each piece has length $\frac{1}{6}$.

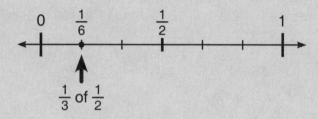

Therefore, $\frac{1}{3}$ of $\frac{1}{2}$ is $\frac{1}{6}$.

PRACTICE | Answer each question below. You may find the number lines helpful.

92. What number is $\frac{1}{5}$ of $\frac{1}{2}$?

93. What number is $\frac{1}{3}$ of $\frac{1}{4}$?

92. _____

93. _____

94. What is $\frac{1}{2}$ of $\frac{1}{3}$?

95. What is $\frac{1}{4}$ of $\frac{1}{2}$?

94. _____

95. _____

Products

EXAMPLE | Compute $\frac{1}{3} \cdot \frac{1}{2}$.

We consider the area model of multiplication. We start with a 1-unit square. We split the square into halves with a vertical line and into thirds with two horizontal lines.

The area of the small shaded rectangle is $\frac{1}{3} \cdot \frac{1}{2}$ square units.

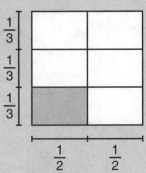

The area of the large square is 1 square unit. Since the square is divided into 6 equal pieces, each piece is one sixth of the area of the large square. So, the area of the small shaded rectangle is $\frac{1}{6}$ square units.

Therefore, $\frac{1}{3} \cdot \frac{1}{2}$ is $\frac{1}{6}$.

In general, we have

$$\frac{1}{a} \cdot \frac{1}{b} = \frac{1}{a \cdot b}.$$

PRACTICE | Compute each product below. You may find the unit squares helpful.

96. Compute $\frac{1}{2} \cdot \frac{1}{5}$.

97. Compute $\frac{1}{4} \cdot \frac{1}{3}$.

96. _____

97. _____

98. Compute $\frac{1}{3} \cdot \frac{1}{6}$.

99. Compute $\frac{1}{3} \cdot \frac{1}{4}$.

98. _____

99. _____

EXAMPLE | Compute $\frac{3}{4} \cdot \frac{1}{2}$.

We know that **one** fourth of $\frac{1}{2}$ is $\frac{1}{8}$.

Three fourths of $\frac{1}{2}$ is **three** times as much.

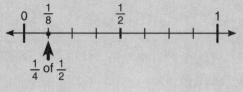

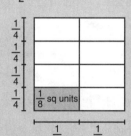

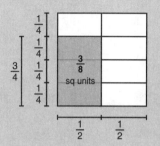

— *or* —

Using the commutative and associative properties of multiplication, we have

$$\frac{3}{4} \cdot \frac{1}{2} = \left(3 \cdot \frac{1}{4}\right) \cdot \frac{1}{2}$$
$$= 3 \cdot \left(\frac{1}{4} \cdot \frac{1}{2}\right)$$
$$= 3 \cdot \frac{1}{8}$$
$$= \frac{3}{8}.$$

PRACTICE | Compute each product below.

100. $\frac{2}{3} \cdot \frac{1}{5} =$

101. $\frac{9}{10} \cdot \frac{1}{10} =$

102. $\frac{1}{2} \cdot \frac{7}{11} =$

103. $\frac{1}{7} \cdot \frac{3}{4} =$

104. $\frac{3}{5} \cdot \frac{1}{4} =$

105. $\frac{1}{9} \cdot \frac{2}{3} =$

106. $\frac{7}{8} \cdot \frac{1}{5} =$

107. $\frac{5}{6} \cdot \frac{1}{3} =$

EXAMPLE | Compute $\frac{5}{8} \cdot \frac{7}{9}$.

We use the commutative and associative properties of multiplication to compute this product.

We have $\frac{5}{8} \cdot \frac{7}{9} = \left(5 \cdot \frac{1}{8}\right) \cdot \left(7 \cdot \frac{1}{9}\right)$

$$= (5 \cdot 7) \cdot \left(\frac{1}{8} \cdot \frac{1}{9}\right)$$

$$= 35 \cdot \frac{1}{72}$$

$$= \frac{35}{72}.$$

To multiply two fractions, we multiply their numerators to get the numerator of the product, and we multiply their denominators to get the denominator of the product.

In general, we have
$$\frac{a}{b} \cdot \frac{c}{d} = \frac{a \cdot c}{b \cdot d}.$$

PRACTICE | Compute each product below.

108. $\frac{7}{10} \cdot \frac{3}{4} =$

109. $\frac{2}{3} \cdot \frac{5}{7} =$

110. $\frac{2}{13} \cdot \frac{3}{5} =$

111. $\frac{2}{7} \cdot \frac{4}{7} =$

112. $\frac{8}{9} \cdot \frac{4}{5} =$

113. $\frac{5}{8} \cdot \frac{9}{11} =$

114. $\frac{3}{4} \cdot \frac{7}{11} =$

115. $\frac{5}{6} \cdot \frac{5}{9} =$

EXAMPLE | Compute $\frac{14}{15} \cdot \frac{4}{35}$.

$\frac{14}{15} \cdot \frac{4}{35} = \frac{14 \cdot 4}{15 \cdot 35} = \frac{56}{525}$, which simplifies to $\frac{8}{75}$.

— *or* —

Both 14 and 35 are multiples of 7, so we rearrange the multiplication as shown:

$$\frac{14}{15} \cdot \frac{4}{35} = \frac{14 \cdot 4}{15 \cdot 35} = \frac{4 \cdot 14}{15 \cdot 35} = \frac{4}{15} \cdot \frac{14}{35}.$$

Then, we divide both 14 and 35 by their common factor of 7 to simplify the expression:

$$\frac{4}{15} \cdot \frac{14}{35} = \frac{4}{15} \cdot \frac{2}{5} = \frac{8}{75}.$$

As a shortcut, we can begin by dividing both 14 and 35 by their GCF (7). We write our work as shown below:

$$\frac{{}^{2}\cancel{14}}{15} \cdot \frac{4}{\cancel{35}_{5}} = \frac{2 \cdot 4}{15 \cdot 5} = \frac{8}{75}.$$

Simplifying first can make a computation easier.

Dividing numbers in the numerator and denominator by common factors is often called **cancelling**.

PRACTICE | Compute each product below.

116. $\frac{2}{3} \cdot \frac{1}{2} =$

117. $\frac{5}{6} \cdot \frac{6}{11} =$

118. $\frac{2}{3} \cdot \frac{5}{8} =$

119. $\frac{4}{5} \cdot \frac{3}{14} =$

120. $\frac{3}{10} \cdot \frac{4}{9} =$

121. $\frac{9}{10} \cdot \frac{8}{15} =$

122. $\frac{3}{8} \cdot \frac{5}{6} \cdot \frac{4}{5} =$

123. ★ $\frac{2}{5} \cdot \frac{7}{10} \cdot \frac{20}{3} \cdot \frac{5}{21} \cdot \frac{9}{2} =$

PRACTICE | Fill in each blank below to make a true statement. ***All fractions must be less than 1 and in simplest form.***

124. $\dfrac{}{} \cdot \dfrac{6}{7} = \dfrac{5}{14}$

125. $\dfrac{2}{27} \cdot \dfrac{}{40} = \dfrac{1}{60}$

126. $\dfrac{5}{36} \cdot \dfrac{12}{} = \dfrac{1}{15}$

127. $\dfrac{}{24} \cdot \dfrac{8}{65} = \dfrac{1}{39}$

128. $\dfrac{}{24} \cdot \dfrac{}{40} = \dfrac{3}{64}$

129. $\dfrac{}{16} \cdot \dfrac{}{33} = \dfrac{1}{3}$

130. ★ $\dfrac{}{32} \cdot \dfrac{20}{} = \dfrac{45}{56}$

131. ★ $\dfrac{8}{} \cdot \dfrac{}{10} = \dfrac{12}{25}$

EXAMPLE | Compute $10\frac{1}{3} \cdot \frac{2}{5}$.

We first convert the mixed number into an equivalent fraction, then multiply the two fractions:

$10\frac{1}{3} = \frac{31}{3}$, so

$10\frac{1}{3} \cdot \frac{2}{5} = \frac{31}{3} \cdot \frac{2}{5} = \frac{62}{15} = 4\frac{2}{15}$.

— *or* —

We use the distributive property:

$$10\frac{1}{3} \cdot \frac{2}{5} = \left(10 + \frac{1}{3}\right) \cdot \frac{2}{5}$$
$$= \left(10 \cdot \frac{2}{5}\right) + \left(\frac{1}{3} \cdot \frac{2}{5}\right)$$
$$= \frac{20}{5} + \frac{2}{15}$$
$$= 4 + \frac{2}{15}$$
$$= 4\frac{2}{15}$$
$$= \frac{62}{15}.$$

So, $10\frac{1}{3} \cdot \frac{2}{5} = \frac{62}{15} = 4\frac{2}{15}$.

PRACTICE | Compute each product below.

132. $1\frac{3}{5} \cdot \frac{1}{2} =$

133. $\frac{5}{8} \cdot 1\frac{1}{4} =$

134. $2\frac{2}{3} \cdot \frac{3}{4} =$

135. $\frac{4}{5} \cdot 3\frac{1}{8} =$

136. $\frac{1}{6} \cdot 9\frac{6}{7} =$

137. $2\frac{1}{7} \cdot 4\frac{1}{5} =$

138. $1\frac{7}{9} \cdot 2\frac{1}{10} =$

139. $8\frac{9}{14} \cdot \frac{7}{8} =$

PRACTICE | Answer each question below.

140. In Erin's class, $\frac{4}{7}$ of the students have pets, and $\frac{3}{8}$ of the students who have pets have dogs. What fraction of the students in Erin's class have dogs?

140. _____

141. What is the area of a square with side length $2\frac{1}{10}$ meters?

141. _____

142. Suppose a and b are two fractions between 0 and 1, and a is less than b. Order a, b, and the product ab from least to greatest.

142. _____ < _____ < _____

143. ★ On Wednesday, Tim read $\frac{1}{4}$ of his book. On Thursday, he read $\frac{2}{3}$ of what was left. On Friday, he finished the book. What fraction of the book did Tim read on Friday?

143. _____

144. ★★ Amy gives $\frac{2}{5}$ of her pennies to Ben, and then gives $\frac{3}{4}$ of her remaining pennies to Chandra. Amy then uses her twelve remaining pennies to make wishes in a fountain. How many pennies did Amy have before she gave some to Ben?

144. _____

EXAMPLE | Use the given numbers to fill in the blanks to create a true equation. *All fractions must be less than 1 and in simplest form.*

Numbers: 1, 5, 15

$$\frac{}{6} \cdot \frac{2}{} = \frac{}{9}$$

We first place the 15. Since all fractions are less than 1 and in simplest form, 15 can only be the denominator of $\frac{2}{}$.

$$\frac{}{6} \cdot \frac{2}{15} = \frac{}{9}$$

2 and 6 have a greatest common factor of 2, so the 2 and 6 cancel as shown.

$$\frac{}{\cancel{6}_3} \cdot \frac{\cancel{2}^1}{15} = \frac{}{9}$$

The denominator of the product is $9 = 3 \cdot 3$, and $15 = 3 \cdot 5$. So, the numerator of $\frac{}{6}$ must cancel the factor of 5 in the denominator of $\frac{}{15}$.

$$\frac{5}{\cancel{6}_3} \cdot \frac{\cancel{2}^1}{15} = \frac{}{9}$$

Between the remaining numbers 1 and 5, only 5 is a multiple of 5. So, 5 is the numerator of $\frac{}{6}$.

Finally, 1 fills the remaining blank.

$$\frac{5}{6} \cdot \frac{2}{15} = \frac{1}{9}$$

We check that the equation is true:

$$\frac{5}{6} \cdot \frac{2}{15} = \frac{\cancel{5}^1}{\cancel{6}_3} \cdot \frac{\cancel{2}^1}{\cancel{15}_3} = \frac{1}{9}. \quad \checkmark$$

PRACTICE | For each problem below, use the given numbers to fill in the blanks to create a true equation. *All fractions must be less than 1 and in simplest form.*

145. Numbers: 10, 11, 25

$$\frac{21}{} \cdot \frac{}{} = \frac{42}{55}$$

146. Numbers: 12, 15, 22

$$\frac{11}{} \cdot \frac{}{} = \frac{5}{8}$$

147. Numbers: 3, 5, 6

$$\frac{4}{} \cdot \frac{}{10} = \frac{}{25}$$

148. Numbers: 3, 4, 5

$$\frac{5}{6} \cdot \frac{}{} = \frac{2}{}$$

Fill-In Products

PRACTICE | For each problem below, use the given numbers to fill in the blanks to create a true equation. ***All fractions must be less than 1 and in simplest form.***

149. Numbers: 3, 7, 14

$$\frac{9}{\underline{}} \cdot \frac{\underline{}}{12} = \frac{\underline{}}{8}$$

150. Numbers: 5, 8, 24, 25

$$\frac{\underline{}}{\underline{}} \cdot \frac{5}{\underline{}} = \frac{3}{\underline{}}$$

151. Numbers: 3, 9, 15, 22

$$\frac{\underline{}}{\underline{}} \cdot \frac{11}{\underline{}} = \frac{\underline{}}{10}$$

152. Numbers: 4, 7, 9, 14

$$\frac{\underline{}}{18} \cdot \frac{\underline{}}{\underline{}} = \frac{1}{\underline{}}$$

153. Numbers: 4, 10, 14, 15, 21

$$\frac{\underline{}}{\underline{}} \cdot \frac{\underline{}}{\underline{}} = \frac{\underline{}}{9}$$

154. Numbers: 2, 5, 6, 7, 21

$$\frac{\underline{}}{\underline{}} \cdot \frac{\underline{}}{\underline{}} = \frac{5}{\underline{}}$$

155. Numbers: 2, 3, 5, 9, 10

$$\frac{\underline{}}{\underline{}} \cdot \frac{\underline{}}{\underline{}} = \frac{3}{\underline{}}$$

156. Numbers: 6, 10, 12, 25

$$\frac{5}{\underline{}} \cdot \frac{\underline{}}{\underline{}} = \frac{1}{\underline{}}$$

The **reciprocal** of a number n is the number we multiply by n to get 1.

EXAMPLE | What is the reciprocal of 55?

$55 \cdot \boxed{\frac{1}{55}} = 1$, so the reciprocal of 55 is $\frac{1}{55}$.

Review the introduction to reciprocals in Chapter 10 of Beast Academy 4D!

PRACTICE | Write the reciprocal of each number in simplest form.

157. $\frac{1}{7}$ Reciprocal: _____

158. 25   Reciprocal: _____

159. $2+8$ Reciprocal: _____

160. $9 \cdot 4$ Reciprocal: _____

161. $\frac{24}{6}$ Reciprocal: _____

162. $\frac{11}{55}$ Reciprocal: _____

PRACTICE | Answer each question below.

To divide by a number, we can multiply by its **reciprocal**.

163. $7 \div \frac{1}{5} =$

164. $5 \div \frac{1}{6} =$

165. $2 \div \left(5 \div \frac{1}{8}\right) =$

166. $3 \div \left(\frac{1}{4} \div \frac{1}{2}\right) =$

167. If $a \div \frac{1}{23} = 17$, what is the value of a?   167. _____

The *reciprocal* of a number n is the number we multiply by n to get 1.

EXAMPLE | What is the reciprocal of $\frac{2}{5}$?

$\frac{2}{5} \cdot \boxed{\frac{5}{2}} = \frac{10}{10} = 1$, so the reciprocal of $\frac{2}{5}$ is $\frac{5}{2}$.

In general, we have $\frac{a}{b} \cdot \frac{b}{a} = \frac{ab}{ab} = 1$.

So, as long as a and b are not zero, we can find the reciprocal of any fraction $\frac{a}{b}$ by switching the numerator and denominator: $\frac{b}{a}$.

PRACTICE | Write the reciprocal of each number in simplest form.

168. $\frac{2}{3}$ Reciprocal: _____

169. $\frac{3}{8}$ Reciprocal: _____

170. $2\frac{1}{2}$ Reciprocal: _____

171. $5\frac{1}{3}$ Reciprocal: _____

172. $2\frac{2}{5}$ Reciprocal: _____

173. $4 \cdot \frac{5}{9}$ Reciprocal: _____

174. $\frac{4}{7} \cdot \frac{3}{5}$ Reciprocal: _____

175. $\frac{2}{5} + \frac{2}{9}$ Reciprocal: _____

176. What is the *reciprocal* of the sum of one half, one third, and one fourth?

176. _____

EXAMPLE | Compute $\frac{2}{7} \div \frac{3}{5}$.

To divide $\frac{2}{7}$ by $\frac{3}{5}$, we multiply $\frac{2}{7}$ by the reciprocal of $\frac{3}{5}$.

So, $\frac{2}{7} \div \frac{3}{5} = \frac{2}{7} \cdot \frac{5}{3} = \frac{\mathbf{10}}{\mathbf{21}}$.

We check:

$$\frac{3}{5} \cdot \frac{10}{21} = \frac{\overset{1}{\cancel{3}}}{\underset{1}{\cancel{5}}} \cdot \frac{\overset{2}{\cancel{10}}}{\underset{7}{\cancel{21}}} = \frac{2}{7}. \checkmark$$

PRACTICE | Compute each quotient below.

177. $9 \div \frac{3}{4} =$

178. $\frac{5}{8} \div \frac{7}{9} =$

179. $\frac{1}{4} \div \frac{3}{8} =$

180. $\frac{5}{6} \div \frac{4}{5} =$

181. $1\frac{5}{8} \div \frac{2}{3} =$

182. $\frac{8}{15} \div \frac{12}{25} =$

183. $\frac{21}{16} \div \frac{49}{50} =$

184. $\frac{52}{55} \div 2\frac{6}{11} =$

185. $\frac{6}{35} \div \left(\frac{3}{14} \div \frac{4}{5} \right) =$

186. $\left(\frac{6}{35} \div \frac{3}{14} \right) \div \frac{4}{5} =$

These multiplication tables are just like the tables we used when practicing the multiplication facts from 1 • 1 to 10 • 10.

The number in any white square is the product of the shaded numbers above and to the left of the white square.

EXAMPLE Fill in the missing entries in the multiplication table to the right.

The number in the bottom-left white square is $\frac{3}{7} \cdot \frac{1}{2} = \frac{3}{14}$.

Next, we look for the number in the empty shaded square in the left column.

Since $\boxed{} \cdot \frac{1}{2} = \frac{1}{5}$, we know $\frac{1}{5} \div \frac{1}{2} = \boxed{}$.

$\frac{1}{5} \div \frac{1}{2} = \frac{1}{5} \cdot 2 = \frac{2}{5}$, so the number in this square is $\frac{2}{5}$.

Next, we consider the empty shaded square in the top row.

Since $\boxed{} \cdot \frac{3}{7} = \frac{2}{7}$, we know $\frac{2}{7} \div \frac{3}{7} = \boxed{}$.

$\frac{2}{7} \div \frac{3}{7} = \frac{2}{7} \cdot \frac{7}{3} = \frac{2}{3}$, so the number in this square is $\frac{2}{3}$.

Finally, we complete the puzzle with $\frac{2}{5} \cdot \frac{2}{3} = \frac{4}{15}$ as shown.

Print more multiplication tables at BeastAcademy.com!

PRACTICE Complete each multiplication table below. Write every entry in simplest form.

187.

•		$\frac{3}{5}$	
$\frac{1}{4}$	$\frac{1}{10}$		$\frac{1}{5}$
		$\frac{9}{20}$	

188.

•		$\frac{5}{6}$	
	$\frac{2}{15}$	$\frac{1}{6}$	$\frac{1}{8}$
	$\frac{1}{2}$		

PRACTICE | Complete each multiplication table below.
Write every entry in simplest form.

189.

$\cdot$		$\frac{5}{6}$	
$\frac{5}{8}$			
	$\frac{4}{9}$		$\frac{10}{21}$
$\frac{5}{11}$			$\frac{30}{77}$

190.

$\cdot$	$\frac{2}{3}$	$\frac{7}{8}$	$\frac{2}{5}$
	$\frac{4}{15}$		
		$\frac{7}{24}$	
			$\frac{3}{10}$

191.

$\cdot$		$\frac{1}{6}$	
$\frac{1}{3}$	$\frac{1}{8}$		$\frac{1}{4}$
		$\frac{2}{15}$	
		$\frac{5}{42}$	

192.

$\cdot$			
$\frac{1}{6}$	$\frac{1}{10}$		$\frac{1}{9}$
$\frac{4}{9}$		$\frac{2}{9}$	
			$\frac{1}{2}$

193.

$\cdot$	$\frac{1}{2}$		
$\frac{2}{5}$			$\frac{1}{10}$
$\frac{3}{8}$		$\frac{1}{8}$	
		$\frac{4}{27}$	

194.

$\cdot$	$\frac{1}{2}$		
	$\frac{1}{3}$		$\frac{1}{2}$
$\frac{1}{4}$		$\frac{1}{6}$	$\frac{3}{16}$
			$\frac{5}{8}$

PRACTICE | Answer each question below.

195. Elle cuts an 8-inch string into $\frac{2}{3}$-inch pieces.
How many pieces does she make?

195. _____

196. What is the side length of a regular pentagon
with perimeter $4\frac{5}{8}$ centimeters?

196. _____

197. The rectangle below has an area of $5\frac{1}{3}$ square units.
What is its width?

197. _____

$\frac{4}{5}$ units

198. A $22\frac{1}{2}$-ounce can holds $2\frac{1}{2}$ servings of soup. How many
ounces are in each serving of this soup?

198. _____

199. ★ A recipe for 36 cupakes calls for $\frac{3}{4}$ of a cup of butter. If Rob wants
to make 24 cupcakes, how many cups of butter does he need?

199. _____

200. ★ Grogg eats $\frac{1}{3}$ of a watermelon. The remaining watermelon weighs
$7\frac{3}{4}$ pounds. How much did the watermelon weigh before Grogg
started eating it?

200. _____

PRACTICE | Simplify each expression below.

201. $\dfrac{2}{7} \cdot \dfrac{13}{3} \cdot \dfrac{7}{17} \cdot \dfrac{11}{2} \cdot \dfrac{5}{11} \cdot \dfrac{17}{5} \cdot \dfrac{19}{13} =$

202. $\left(\dfrac{1}{2} \cdot \dfrac{2}{3} \cdot \dfrac{3}{4}\right) \div \left(\dfrac{3}{4} \cdot \dfrac{2}{5} \cdot \dfrac{2}{3}\right) =$

203. $\dfrac{2 \cdot 3 \cdot 4 \cdot 5 \cdot 6 \cdot 7 \cdot 8}{4 \cdot 6 \cdot 8 \cdot 10 \cdot 12 \cdot 14 \cdot 16} =$

204. $\dfrac{2}{3} \cdot 11 \cdot \dfrac{5}{13} \cdot 19 \cdot \dfrac{2}{77} \cdot 26 \cdot \dfrac{1}{95} \cdot 27 =$

205. $\left(\dfrac{5}{13} \cdot \dfrac{8}{21}\right) + \left(\dfrac{5}{13} \cdot \dfrac{13}{21}\right) + \left(\dfrac{5}{13} \cdot \dfrac{17}{21}\right) + \left(\dfrac{5}{13} \cdot \dfrac{4}{21}\right) =$

206. Recall that $n!$ means "n factorial": the product of all positive whole numbers less than or equal to n.

Simplify $\dfrac{99!}{101!}$.

206. _____

PRACTICE | Answer each question below.

207. Rosencrantz and Guildenstern paint lines around the perimeter of a rectangular field that is 120 yards long and 50 yards wide. It takes $6\frac{1}{4}$ cans of spray paint to paint all of the lines. How many yards of line will one full can paint?

207. _____

208. What is the median of the four numbers below?

$$\frac{3}{4}, \frac{5}{6}, \frac{1}{2}, \frac{4}{7}$$

208. _____

209. How much smaller is $\frac{100}{101}$ than its reciprocal?

209. _____

210. If $\frac{1}{2}$ of $\frac{2}{3}$ of $\frac{3}{4}$ of Julian's age in years is 9, how old is Julian?
★

210. _____

211. Fill in each denominator below with a number or expression to write it as an Egyptian fraction. Assume that m is not 0.

a. $\dfrac{1}{1} = \dfrac{1}{\rule{1cm}{0.4pt}} + \dfrac{1}{\rule{1cm}{0.4pt}} + \dfrac{1}{\rule{1cm}{0.4pt}}$

b. ★ $\dfrac{1}{m} = \dfrac{1}{\rule{1cm}{0.4pt}} + \dfrac{1}{\rule{1cm}{0.4pt}} + \dfrac{1}{\rule{1cm}{0.4pt}}$

c. $\dfrac{1}{7} = \dfrac{1}{\rule{1cm}{0.4pt}} + \dfrac{1}{\rule{1cm}{0.4pt}} + \dfrac{1}{\rule{1cm}{0.4pt}}$

PRACTICE | Answer each question below.

212. The average of $\frac{3}{4}$, $\frac{4}{3}$, and n is 1. What is n?

212.  _____

★

213. Circle the expression below that gives the result of dividing $\frac{a}{b}$ by its reciprocal.

★

$$1 \qquad \frac{a}{b} \qquad \frac{b}{a} \qquad \frac{2a}{2b} \qquad \frac{a^2}{b^2} \qquad \frac{b^2}{a^2}$$

214. Grogg has $\frac{1}{2}$, $\frac{1}{3}$, $\frac{1}{4}$, and $\frac{1}{5}$-gram weights. What is the largest weight less than 1 gram that Grogg can make by combining two or more of these weights?

214.  _____

★

215. If $\frac{2}{3}$ of a tank of gas is enough for Haylie to drive $\frac{3}{4}$ of the way from home to her grandmother's house, what fraction of a tank does Haylie need to drive the whole distance from home to her grandmother's?

215. _____

★

216. A pitcher is $\frac{1}{3}$ full. After Winnie pours out 8 ounces of juice, the pitcher is $\frac{1}{5}$ full. How many ounces of juice are left in the pitcher?

216. _____

★

HINTS
For Selected Problems

Below are hints to every problem marked with a ★.
Work on the problems for a while before looking at the hints.
The hint numbers match the problem numbers.

CHAPTER 4
Statistics   6

18. What happens when both teams have an odd number of students? Both even? One odd and one even?

38. How do all six numbers balance around the desired average, 78?

39. These numbers are all pretty close together. What number could we easily compare these four to?

40. How do Kat's and Matt's scores balance around the team average? What does this tell us about Pat's score?

53. Which of the empty triangles are odd? Which are even?

54. Consider each empty triangle around ⬛6⬛. What numbers could go in each triangle? What is the sum of the numbers in these four triangles?

55. Consider each empty triangle around ⬛5⬛. What numbers could go in each triangle? What is the sum of the numbers in these four triangles?

56. Can you determine what number goes in the triangle that borders the ⬛2⬛ and the ⬛7⬛?

57. Consider the ⬛6⬛ and the ⬛7⬛ at the top of the puzzle. What numbers could go in the empty triangles around each of these squares? From this information, can you determine the number in the top empty square?

58. What number must go in the empty square between the △3△ and the △4△?

76. How does the weight of Trina's flamingoat balance around the Tuesday morning average of 8? How do the other weights balance around the Tuesday morning average?

77. How does the age of Betsy's son balance around the new average?

86. What possible values of x give this list a mode?

87. How large can we make the smallest number?

88. How short can the smallest flower be?

89. What is the height of the tallest monster in class?

96. Consider the right column with average 5 and median 7. What numbers can be used in this column? From this information, can we determine any numbers in the middle row?

97. What numbers can be used in the right column? The left column? How can we make sure the range values are correct?

98. What numbers can be used in the right column? Can you fill in any empty squares using this information?

99. What three digits must go in the top row? What three digits must go in the middle column?

111. Consider the list of all scores from all four students. Can you find the mode(s) of this list? How does this help determine which scores Mr. Erikson left out?

114. Consider each place value in all six arrangements. What is the average digit in each place value?

115. Remember that each bridge has 2 ends!

116. How does the sum $x+y+z$ relate to the sum of $3x+y$, $3y+z$, and $3z+x$?

117. What does the average of $10\frac{1}{2}$ tell us about the middle number(s) in this list?

118. What are the possible values for the median of this list?

119. What are the possible values of the mode of this list? How about the median?

120. How many times does 21 appear in the list?

CHAPTER 5
Factors & Multiples   32

16. Is there a power of 2 that is a factor of 2,772 but not 4,200? A power of 3? A power of 7? A power of 11?

24. How can we use the prime factorization of 504 to find its largest odd factor?

25. $6 = 2 \cdot 3$. How does this help determine the largest power of 6 Myrtle can divide by to get an integer quotient?

44. What expression represents the GCF of $6r$ and $8r$?

45. How can the GCF of a and b help us find the second-largest factor they have in common?

46. How are the common factors of 1,008 and 1,620 related to the GCF of 1,008 and 1,620?

55. Try different examples for a and b! Smaller numbers are usually easier to work with.

75. What is the smallest number that is divisible by 4, 5, and 6? The second-smallest?

78. What expression represents the LCM of $12p$ and $18p$?

88. How can a Venn diagram help you explain this?

102. One ninth of the gymnasts at Acrobeast Gymnasium have a tail. What does that tell us about the total number of gymnasts? Remember, the number of gymnasts with a tail must be an integer!

103. How many stairs are there between consecutive floors?

104. At the end of the game, the total points Niki *gained* must equal the total points she *lost*.

114. Consider the bottom two blank circles. Their GCF is 1, and their LCM is 72. What numbers could go in these two circles?

115. What do the prime factors of 44 and 396 tell us about the prime factors of the number in the circle between them?

116. What do the prime factors of 26 and 104 tell us about the prime factors of the number in the circle between them?

119. What number must go in the blank circle between 1 and 25?

120. Consider the two bottom blank circles. What two numbers must fill these circles?

124. How can the prime factorization of 6 help us find the largest power of 6 that is a factor of (12!)?

128. What combination of factors cause a number to have trailing zeros?

129. How can you use the strategies learned in Problems 125-128 to solve this problem?

137. Try drawing a Venn diagram (as on Page 48) for two relatively prime numbers. What do you notice?

143. What is the smallest number that is divisible by 6, 9, and 21?

144. Try drawing a Venn diagram for the two numbers. What section(s) can you fill in right away?

145. How many trailing zeros does this number have?

146. What do we know about the prime factorization of a number with 7 trailing zeros?

147. What is the prime factorization of (10!)?

148. What is the LCM of 168 and 980? How can this be used to find the smallest possible value of Lizzie's third number?

149. Try a similar problem with smaller numbers. What if LCM$(x, y) = 12$? What if LCM$(x, y) = 20$? What if LCM$(x, y) = 30$?

150. What are the possible dimensions of each monster's rectangle? Which of these can be connected to create a rectangle?

151. Consider the prime factorization of each number. Which number *must* be circled?

152. Consider each pair of bells separately.

153. How do the coin values balance around the average?

CHAPTER 6
Fractions
64

37. c. How many centimeters did Rosa grow during these 6 months?

38. How much would two hexatoads weigh?

39. Remember to follow the correct order of operations!

40. For what value of C is $\frac{1}{C} + \frac{1}{C} = \frac{1}{3}$? How does this limit the possibilities for A and B?

63. What is the largest fraction that can be part of this sum? The second-largest?

82. Place the 2 first. Then, where does the 3 go?

83. Where can you place the 42? The 28?

123. What factors can you cancel?

130. Don't forget that all fractions must be less than 1!

131. Don't forget that all fractions must be less than 1 *and* in simplest form!

143. What fraction of the *whole* book did Tim read on Thursday?

144. What fraction of the original number of pennies does Amy have after giving pennies to Chandra?

— *or* —

Try working backwards. How many pennies did Amy have before sharing with Chandra?

199. The original recipe makes 36 cupcakes. What fraction of the original recipe is used to make 24 cupcakes?

200. After Grogg eats $\frac{1}{3}$ of the watermelon, what fraction of the watermelon is left? Can you determine what *one* third of the watermelon weighs?

— *or* —

How would you solve a similar problem that did not include fractions? For example, if 4 watermelons weigh 12 pounds, how much does each watermelon weigh?

210. What is $\frac{1}{2}$ of $\frac{2}{3}$ of $\frac{3}{4}$?

211. b. How can we use part (a) to help us solve part (b)?

212. What is the sum of $\frac{3}{4}$, $\frac{4}{3}$, and n?

213. Dividing $\frac{a}{b}$ by its reciprocal is the same as multiplying $\frac{a}{b}$ by what?

214. Write all of the weights with the same denominator. What is the least common denominator that can be used for all of Grogg's weights?

215. How many tanks of gas are required to make it *one* fourth of the way to Haylie's grandmother's house?

— *or* —

How would you solve a similar problem that did not include fractions? For example, if 6 tanks of gas are used to make 3 trips to grandma's, then how many tanks of gas are used to make 1 trip to grandma's?

216. What fraction of the pitcher is the 8 ounces of juice Winnie poured out?

SOLUTIONS
Chapters 4-6

1. Depending on the responses you receive, your answers below will vary.

a. When the people are arranged in order from youngest to oldest, the age of the person in the middle is the *median* age. You will learn more about the median beginning on page 8.

b. The difference in height between the tallest person and shortest person is the *range* of the heights. You will learn more about the range beginning on page 24.

c. The most common integer is the *mode*. If two or more numbers were chosen the most, those are all the modes. If each integer appears only once, then no integer is the most common, and there is no mode. You will learn more about the mode beginning on page 24.

d. We use this sum to help us answer the next question.

e. The number of hours each person gets is the *average* (or *mean*) number of hours of sleep. You will learn more about the average beginning on page 12.

2. We write the eight ages from least to greatest:

8, 8, 8, 8, 8, 9, 9, 9.

The two numbers in the middle of this list are 8 and 8. In this case, we see that the median age is **8 years** without any computation.

3. There are five two-legged monsters. Writing their weights from least to greatest, we have

39, 53, 55, 55, 70.

The median weight of the two-legged monsters is **55 lbs**.

4. There are six monsters who have horns. Writing the horn lengths from least to greatest, we have

2, 5, 7, 11, 12, 15.

The two numbers in the middle are 7 and 11, so the median horn length is the number halfway between those two numbers: $\frac{7+11}{2} = \frac{18}{2} = $ **9 inches**.

5. There are five 8-year-old monsters. Writing their heights from least to greatest, we have

28, 34, 45, 48, 50.

The median height of the 8-year-olds is 45 inches.

There are three 9-year-olds. Writing their heights from least to greatest, we have

18, 30, 32.

The median height of the 9-year-olds is 30 inches.

So, the median height of the 8-year-olds is 45−30 = **15** inches greater than the median height of the 9-year-olds.

6. We first list the number of legs and the number of eyes in order from least to greatest.

Legs: 2, 2, 2, 2, 2, 3, 4, 4.   Eyes: 2, 2, 2, 2, 2, 3, 3, 5.

So, the median number of legs is 2, and the median number of eyes is 2.

In a previous problem, we found that the median age is 8. Nine-year-old **Oksbert** is the only monster with 2 legs and 2 eyes who is *not* 8 years old.

7. Adding one monster to the group in the table gives us an odd number of monsters. So, the median height and weight will be equal to the middle numbers of those lists.

Therefore, for the new monster to join the group without changing the median height or weight, the new monster's height and weight must be equal to the median height and weight of the 8 other monsters.

The heights and weights of the 8 original monsters are listed below in order from least to greatest.

Heights: 18, 28, 30, 32, 34, 45, 48, 50.

Weights: 7, 19, 29, 39, 53, 55, 55, 70.

So, the median height of the 8-monster group is $\frac{32+34}{2} = \frac{66}{2} = 33$ inches, and the median weight is $\frac{39+53}{2} = \frac{92}{2} = 46$ pounds.

Among the monsters who could join this group, we see that only **Mitch** is 33 inches tall and weighs 46 pounds.

Name	Height	Weight
Yosh	33 in	41 lbs
Tildie	32 in	46 lbs
Kres	34 in	41 lbs
Mitch	33 in	46 lbs
Cad	34 in	39 lbs
Erma	32 in	39 lbs

8. We start with the list of heights of the 8 original monsters: 18, 28, 30, 32, 34, 45, 48, 50.

The median height of the 9-monster list is the middle number of the new list.

- If the new monster's height is less than or equal to 32 inches, then the median height is 32. For example,

18, 28, 30, *31*, <u>32</u>, 34, 45, 48, 50.

- If the new monster's height is 33 inches, then the median height is 33:

18, 28, 30, 32, *<u>33</u>*, 34, 45, 48, 50.

- If the new monster's height is greater than or equal to 34 inches, then the median height is 34. For example,

18, 28, 30, 32, <u>34</u>, 45, 48, *49*, 50.

So, the possible median heights are **32, 33, and 34 inches**.

9. There are four ways to split the group of numbers with a vertical or horizontal line.

```
1 | 3  5  7        1  3 | 5  7
2 | 4  6           2  4 | 6

1  3  5 | 7        1  3  5  7
2  4  6 |          ——————————
                   2  4  6
```

Only the horizontal bar below separates the numbers into two groups with the same median: 4.

```
1  3  5  7     Median: 4
——————————
2  4  6        Median: 4
```

The only solution for each puzzle is shown below.

10.

```
9  7 | 10
   3 | 4
Median: 7   Median: 7
```

11.

```
     | 4
   2 | 3  6
7  5 | 8
Median: 5   Median: 5
```

12.

```
1  8  4  6   Median: 5
——————————
5  3  9      Median: 5
```

13.

```
    2  6     Median: 4
——————————
11  3  5     Median: 4
       3
```

14.

```
   1 | 2  4
   9 | 5  5
   4 | 3
Median: 4   Median: 4
```

15. We write the 10 numbers in each puzzle in order from least to greatest to find each median.

a. 1, 2, 3, 4, 5, 5, 7, 7, 8, 9 Median: **5**

b. 2, 3, 4, 4, 5, 7, 7, 8, 8, 9 Median: **6**

c. 1, 1, 2, 3, 3, 4, 5, 6, 8, 8 Median: **$3\frac{1}{2}$**

16. a.

```
       8
1  4  5  9     Median: 5
——————————
   7  7  2
   5  3        Median: 5
```

b.

```
2  7
8  4  5        Median: 6
   4  8  7
——————————
3  9           Median: 6
```

c.

```
   6  4 | 5
3  8  8 |
   1  3 | 2
      1 |
```

Median: $3\frac{1}{2}$ Median: $3\frac{1}{2}$

17. In each of the three puzzles, the median of the ten-number group is **the same as** the median in each divided group in the corresponding puzzle.

18. The number of monsters on the two teams are both odd, one odd and one even, or both even.

- If both teams have an odd number of monsters, then 46 inches is the middle height of both lists. So, the ordered lists of heights look like this:

  __, __, __, ..., 46, ..., __, __, __
  (x heights less than or equal to 46; x heights greater than or equal to 46)

  __, __, __, ..., 46, ..., __, __, __
  (y heights less than or equal to 46; y heights greater than or equal to 46)

  Then, the combination of these two lists has an even number of items and looks like this:

  __, __, __, ..., 46, 46, ..., __, __, __
  ($x+y$ heights less than or equal to 46; $x+y$ heights greater than or equal to 46)

  In this case, the median of the whole class is 46.

  For example, (27, 46, 60) and (4, 44, 46, 75, 80) combine to make (4, 27, 44, 46, 46, 60, 75, 80).

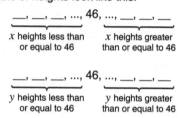

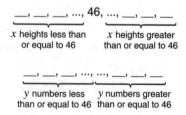

- If one team has an odd number of monsters and the other has an even number, then the ordered lists of heights look like this:

  __, __, __, ..., 46, ..., __, __, __
  (x heights less than or equal to 46; x heights greater than or equal to 46)

  __, __, __, ..., ..., __, __, __
  (y numbers less than or equal to 46; y numbers greater than or equal to 46)

  So, the combination of these two lists has an odd number of items and looks like this:

  __, __, __, ... , 46, ... , __, __, __
  ($x+y$ heights less than or equal to 46; $x+y$ heights greater than or equal to 46)

Therefore, the median of the whole class is 46.

For example, (4, 45, 47, 88) and (9, 46, 60) combine to make (4, 9, 45, <u>46</u>, 47, 60, 88).

- If both teams have an even number of monsters, then 46 inches is the height halfway between the two middle heights for each team.

We call the heights of the middle two monsters on the first team a and b, and we call the heights of the middle two monsters on the second team c and d.

Then, the ordered lists of heights look like this:

$$\underbrace{\underline{}, \underline{}, \underline{}, ...,}_{\substack{x \text{ heights less than} \\ \text{or equal to } a}} a, b, \underbrace{..., \underline{}, \underline{}, \underline{}}_{\substack{x \text{ heights greater} \\ \text{than or equal to } b}}$$

$$\underbrace{\underline{}, \underline{}, \underline{}, ...,}_{\substack{y \text{ heights less than} \\ \text{or equal to } c}} c, d, \underbrace{..., \underline{}, \underline{}, \underline{}}_{\substack{y \text{ heights greater} \\ \text{than or equal to } d}}$$

Because the median height of the first team is 46 inches, we know that a and b are exactly the same distance from 46: a is less than 46 by the same amount that b is greater than 46.

Similarly, on the second team, c and d are exactly the same distance from 46, with c less than 46 by the same amount that d is greater than 46.

If a and b are closer to 46 than c and d are, then we have

$$\underbrace{\underline{}, \underline{}, \underline{}, ...,}_{\substack{x+y+1 \text{ heights less} \\ \text{than or equal to } a \\ (\text{including } c)}} a, b, \underbrace{..., \underline{}, \underline{}, \underline{}}_{\substack{x+y+1 \text{ heights greater} \\ \text{than or equal to } b \\ (\text{including } d)}}$$

Otherwise, c and d are closer to 46 than a and b are, and we have

$$\underbrace{\underline{}, \underline{}, \underline{}, ...,}_{\substack{x+y+1 \text{ heights less} \\ \text{than or equal to } c \\ (\text{including } a)}} c, d, \underbrace{..., \underline{}, \underline{}, \underline{}}_{\substack{x+y+1 \text{ heights greater} \\ \text{than or equal to } d \\ (\text{including } b)}}$$

Whichever team's middle monsters are closer to 46 inches will also be the two middle monsters for the whole class. In the case of a tie, we could pick either pair to be the middle.

So, the median height for the whole class is 46 inches.

For example, (4, 44, 48, 61) and (10, 40, 52, 66) combine to make (4, 10, 40, <u>44</u>, <u>48</u>, 52, 61, 66).

In every possible case, the median height of the whole class is the same as the median of both teams: 46 inches.

*Something to think about: If the two teams' heights **do not** have the same median, is the median of the whole class equal to the number halfway between the medians of the two teams?*

19. We consider a sample list with a median of 10:

9, 10, 10, 11, 15.

We can split this list into the two groups below:

(9, 10, 10), and (11, 15).

The median of the group on the left is 10, but the median of the group on the right is 13.

Yes, it is possible for a group of numbers with a median of 10 to be split into two groups so that one group has a median of 10 but the other group has a different median.

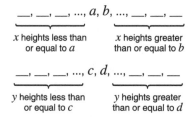

STATISTICS
Average   12-15

20. The little monsters ate a total of $6+7+10+33=56$ marshmallows. So, if the marshmallows were divided equally among the four monsters, each monster would receive $56 \div 4 = $ **14** marshmallows.

21. Alice has a total of $9+13+15+17+16+11+24=105$ dolls. So, if the dolls are divided equally among the seven shelves, each shelf will have $105 \div 7 = $ **15** dolls.

22. The monsters are paid a total of $20 \cdot 5 + 32 = 132$ dollars. After they share the money equally, each of the six monsters has $132 \div 6 = $ **22 dollars**.

23. Mrs. Ryan has $7+3+14+11+18+2+1+0=56$ pencils. If Mrs. Ryan divides the pencils equally among the eight cups, each cup will contain $56 \div 8 = 7$ pencils. So, the cup with three pencils will gain $7-3=$ **4** pencils.

24. $\frac{2+3+5+7+8+11+13}{7} = \frac{49}{7} = $ **7**.

25. $\frac{5+10+20+20+25+35+50+75}{8} = \frac{240}{8} = $ **30**.

— *or* —

Each number is a multiple of 5:

$5 \cdot 1, \ 5 \cdot 2, \ 5 \cdot 4, \ 5 \cdot 4, \ 5 \cdot 5, \ 5 \cdot 7, \ 5 \cdot 10, \ 5 \cdot 15.$

So, we have

$$\frac{5+10+20+20+25+35+50+75}{8} = \frac{5 \cdot (1+2+4+4+5+7+10+15)}{8}$$
$$= \frac{5 \cdot 48}{8}$$
$$= 5 \cdot \frac{48}{8}$$
$$= 5 \cdot 6$$
$$= \mathbf{30}.$$

26. $\frac{-20+(-15)+(-13)+(-9)+(-8)+(-6)+(-6)}{7} = \frac{-77}{7} = $ **-11**.

27. $\frac{9.1+9.6+9.9+10.5+10.9}{5} = \frac{50}{5} = $ **10**.

28. Six numbers with an average of 13 have a sum of $6 \cdot 13 = 78$. If a 2 is added to the list, then the sum of all seven numbers is $78 + 2 = 80$, and their average is $\frac{80}{7} = \mathbf{11\frac{3}{7}}$.

29. Nine numbers with an average of 10 have a sum of $9 \cdot 10 = 90$. If 2 is removed from the list, then the sum of the remaining eight numbers is $90 - 2 = 88$, and their average is $\frac{88}{8} = $ **11**.

30. Four numbers with an average of 40 have a sum of $4 \cdot 40 = 160$. The sum of the three known numbers is $39 + 35 + 44 = 118$. So, the missing number is $160 - 118 = $ **42**.

— *or* —

We consider how the numbers "balance" around the average, 40.

39 and 35 are below the average by a total of 6, and 44 is above the average by 4. So, the three given numbers are a total of 2 below the average.

$$\begin{array}{ccc} & -2 & \\ -1 & -5 & +4 \\ 39 & 35 & 44 \end{array}$$

To balance all of the below-average numbers, we need a number that is above the average by 2. So, the missing number is $40+2=$ **42**.

$$\begin{array}{cccc} -1 & -5 & +4 & +2 \\ 39 & 35 & 44 & \textbf{42} \end{array}$$

31. The four given numbers are a total of 4 below the average, 20.

$$\begin{array}{cccc} & -4 & & \\ -1 & -6 & +5 & -2 \\ 19 & 14 & 25 & 18 \end{array}$$

To balance all of the below-average numbers, we need a number that is above the average by 4. So, the missing number is $20+4=$ **24**.

$$\begin{array}{ccccc} -1 & -6 & +5 & -2 & +4 \\ 19 & 14 & 25 & 18 & \textbf{24} \end{array}$$

32. The four given numbers are a total of 5 above the average, 54.

$$\begin{array}{cccc} & +5 & & \\ +1 & -4 & +2 & +6 \\ 55 & 50 & 56 & 60 \end{array}$$

To balance all of the above-average numbers, we need a number that is below the average by 5. So, the missing number is $54-5=$ **49**.

$$\begin{array}{ccccc} +1 & -4 & +2 & +6 & -5 \\ 55 & 50 & 56 & 60 & \textbf{49} \end{array}$$

33. The four given numbers are a total of 5 above the average, 105.

$$\begin{array}{cccc} & +5 & & \\ +6 & -4 & -2 & +5 \\ 111 & 101 & 103 & 110 \end{array}$$

To balance all of the above-average numbers, we need a number that is below the average by 5. So, the missing number is $105-5=$ **100**.

$$\begin{array}{ccccc} +6 & -4 & -2 & +5 & -5 \\ 111 & 101 & 103 & 110 & \textbf{100} \end{array}$$

34. The five given numbers are a total of 0 above the average, 91.

$$\begin{array}{ccccc} & & +0 & & \\ +4 & -3 & -5 & +6 & -2 \\ 95 & 88 & 86 & 97 & 89 \end{array}$$

So, we do not need to balance the above- or below-average numbers. The missing number is equal to the average: **91**.

$$\begin{array}{cccccc} +4 & -3 & -5 & +6 & -2 & +0 \\ 95 & 88 & 86 & 97 & 89 & \textbf{91} \end{array}$$

35. The five given numbers are a total of 5 below the average, 176.

$$\begin{array}{ccccc} & & -5 & & \\ -2 & 0 & +4 & -1 & -6 \\ 174 & 176 & 180 & 175 & 170 \end{array}$$

To balance all of the below-average numbers, we need a number that is above the average by 5. So, the missing number is $176+5=$ **181**.

$$\begin{array}{cccccc} -2 & 0 & +4 & -1 & -6 & +5 \\ 174 & 176 & 180 & 175 & 170 & \textbf{181} \end{array}$$

36. The first five games were a total of 3 above the final average, 34.

$$\begin{array}{ccccc} & & +3 & & \\ -1 & +4 & -2 & -4 & +6 \\ 33 & 38 & 32 & 30 & 40 \end{array}$$

So, in their sixth game, Orange Academy scored 3 points below the average: $34-3=$ **31** points.

37. Teddy's first six test scores are a total of 6 points below his desired average, 90.

$$\begin{array}{cccccc} & & -6 & & & \\ -9 & +6 & +10 & -2 & +1 & -12 \\ 81 & 96 & 100 & 88 & 91 & 78 \end{array}$$

So, on Teddy's next test, he needs to score 6 points higher than his desired average: $90+6=$ **96**.

38. We consider the difference between each number and the desired average, 78.

$$\begin{array}{cccccc} & & -7 & & & \\ -7 & -5 & -4 & 0 & +3 & +6 \\ 71 & 73 & 74 & 78 & 81 & 84 \end{array}$$

This group is a total of 7 below the desired average.

We are asked to remove *two* numbers, and only **71 and 78** are a total of 7 below the desired average. We remove these two so that the remaining numbers balance around the desired average, 78.

39. Each number is equal to or a little more than 8,950.

$$\begin{array}{cccc} & & +12 & \\ +3 & +2 & +0 & +7 \\ 8{,}953 & 8{,}952 & 8{,}950 & 8{,}957 \end{array}$$

The *sum* of these four numbers is 12 more than the sum would be if the average were equal to 8,950. So, the *average* of these four numbers is $\frac{12}{4}=3$ more than 8,950. Their average is $8{,}950+3=$ **8,953**.

We could also show this as follows:

$$\begin{aligned} \frac{8{,}953+8{,}952+8{,}950+8{,}957}{4} &= \frac{8{,}950 \cdot 4 + (3+2+0+7)}{4} \\ &= \frac{8{,}950 \cdot 4 + 12}{4} \\ &= \frac{8{,}950 \cdot 4 + 3 \cdot 4}{4} \\ &= \frac{(8{,}950+3) \cdot 4}{4} \\ &= \textbf{8{,}953}. \end{aligned}$$

40. We consider the way each person's score balances around the team average.

$$\begin{array}{ccc} & 0 & \\ -5 & +8 & ? \\ \rule{1cm}{0.4pt} & \rule{1cm}{0.4pt} & 82 \\ \text{(Kat)} & \text{(Matt)} & \text{(Pat)} \end{array}$$

Kat's and Matt's scores are a total of 3 above the team average. Pat's score must be 3 below the team average so that the average of these three scores is equal to the team average.

$$\begin{array}{ccc} & 0 & \\ -5 & +8 & -3 \\ \rule{1cm}{0.4pt} & \rule{1cm}{0.4pt} & 82 \\ \text{(Kat)} & \text{(Matt)} & \text{(Pat)} \end{array}$$

Therefore, the team average is 3 more than Pat's score: $82+3=$ **85** points.

41. The average of the four digits
around the square is $\frac{9+3+7+5}{4}=\frac{24}{4}=6$.

42. The average of the four digits around
the square is 5, so their sum is $4\cdot5=20$.
Therefore, the digit in the blank triangle
is $20-8-2-4=6$.

43. The average of the four digits around the square is 2,
so their sum is $4\cdot2=8$. Therefore, the sum of the digits
in the two blank triangles is $8-5-1=2$. Each triangle
contains a positive digit, so the digit in each blank
triangle is 1.

44. Since the number in the square is a positive digit, the
sum of the four digits around the square must be divisible
by 4.

$7+5+1=13$. Only a 3 or a 7 could be added to this sum
to get a multiple of 4.

However, if we fill the blank triangle with a 7, the average
of the digits in the four triangles around the square is

$\frac{7+5+1+7}{4}=\frac{20}{4}=5$. Adjacent shapes may not contain the
same digit, so we cannot place a 5 in the square.

After placing a 3 in the blank triangle, the average of the
digits in the four triangles is $\frac{7+5+1+3}{4}=\frac{16}{4}=4$.

*We use the reasoning discussed in the previous
problems to solve the problems that follow.*

45.   **46.**

47. The sum of the digits in the two triangles around 2 is
$2\cdot2=4$. Adjacent shapes may not contain the same digit,
so neither triangle can contain a 2. The only other pair
of digits whose sum is 4 is $1+3$. So, these two triangles
contain a 1 and a 3.

One of these triangles also touches 6. If this triangle
contains a 1, then the other triangle that touches 6 must
be $12-1=11$. However, 11 is not a digit.

Therefore, the triangle that touches both 2 and 6
contains a 3, and we complete the puzzle as shown.

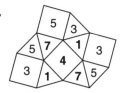

*We use the reasoning discussed in the previous
problems to solve the problems that follow.*

48.   **49.**

50.

51. Step 1:   **Step 2:**   **Final:**

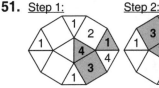

52. Step 1:   **Step 2:**   **Final:**

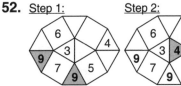

53. The digit in the shaded triangle
that touches 1 and 2 is at
least 3.

At least 3
At least 3, odd

The two digits in the triangles that
touch the shaded square must
have a sum that is divisible by 2.
One of these digits is 1, so the
other is odd and not 1. So, the digit
in the empty triangle is at least 3.

The sum of the digits in the three
triangles around 4 is 12. So, the
digit in the shaded triangle cannot
be more than $12-3-3=6$.

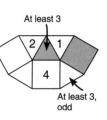
At least 3
Not more than 6
At least 3, odd

The two triangles that touch the
top-left shaded square must have
digits whose sum is divisible by 2.
Since one of these digits is 2, the
other must be even and not 2.
Also, since this empty triangle
touches a 4, it cannot contain a 4.
So the digit in the empty triangle
is 6, and the digit in the shaded
square is 4.

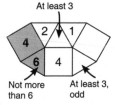
At least 3
Not more than 6
At least 3, odd

Then, the other two triangles that touch the shaded 4 both contain 3's. We complete the puzzle as shown.

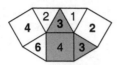

54. The two triangles that touch 2 must contain a 1 and a 3.

The two triangles that touch 3 have a sum of 6, so they contain either a 1 and a 5 or a 2 and a 4.

The two triangles that touch 4 have a sum of 8, so neither can contain a digit greater than 7.

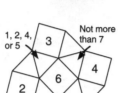

The sum of the four digits around 6 is 24. The largest digit we can place in the shaded triangle is 9, so the sum of the other three digits must be at least 24−9 = 15.

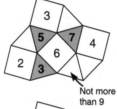

The only way to get a sum of 15 or more with the other three digits around 6 is to choose the largest possible digit for each: 7+5+3 = 15.

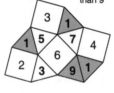

Then, we complete the puzzle as shown.

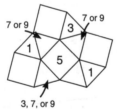

55. We use the reasoning discussed in previous problems to determine the possible digits in the triangles that touch 5.

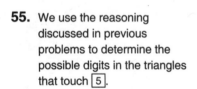

The sum of the digits in the four triangles around 5 is 20. The shaded triangle to the right touches a △1. So, the digit in the shaded triangle is at least 2, and the sum of the other three digits around 5 is no more than 20−2 = 18.

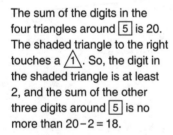

The only way to make a sum of 18 or less in the other three triangles around 5 is to choose the smallest possible digit for each: 3+7+7 = 17.

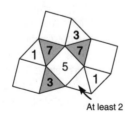

Then, we complete the puzzle as shown.

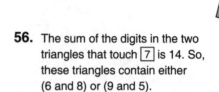

56. The sum of the digits in the two triangles that touch 7 is 14. So, these triangles contain either (6 and 8) or (9 and 5).

Only placing a 5 in the triangle above the 2 as shown allows us to place positive digits in each triangle around 2 to get a sum of 8.

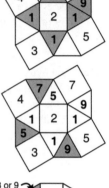

We complete the puzzle as shown.

57. The sum of the digits in the two blank triangles around the shaded 6 is 18−1 = 17. The only way to get a sum of 17 from two digits is 8+9.

The sum of the digits in the two blank triangles around the shaded 7 is 21−4 = 17. The only way to get a sum of 17 from two digits is 8+9.

The sum of the digits in the three triangles around the shaded square must be a multiple of 3. Among our choices, only placing 8's in both blank triangles gives us a sum that is divisible by three: 2+8+8 = 18.

Then, we fill in some of the remaining shapes as shown.

The sum of the digits in the two blank triangles around the shaded 6 is 18−1 = 17. The only way to get a sum of 17 with two digits is 8+9. We place the 8 and 9 as shown.

The sum of the digits in the two blank triangles around the shaded 6 is 18−4 = 14. There are three ways to make a sum of 14 with two digits: 7+7, 6+8, and 5+9.

However, neither triangle can contain a 6, and the triangles can't both contain 7's. This leaves 5 and 9, which we can only place as shown.

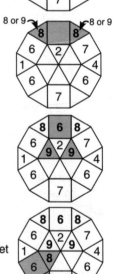

Then, we complete the puzzle as shown.

58. In each of the following steps, there is only one digit that can fill in the shaded triangle so that the sum of the digits around the shaded square is divisible by the number of triangles around it *and* so that touching shapes do not contain the same digit.

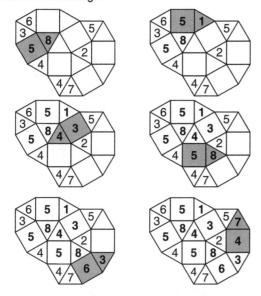

59. The average is $\frac{1+2+3+4+1,000}{5} = \frac{1,010}{5} =$ **202**.

60. The sum of nineteen 2's and 182 is $19 \cdot 2 + 182 = 38 + 182 = 220$.

   So, the average of all 20 numbers is $\frac{220}{20} =$ **11**.

61. The average wage of all 5 employees is $\frac{4 \cdot 10 + 45}{5} = \frac{85}{5} =$ **17 dollars**.

62. The average score on all 5 assignments is $\frac{4 \cdot 100 + 0}{5} = \frac{400}{5} =$ **80 points**.

63. In problem 59, if we exclude the extreme value 1,000, the average value of (1, 2, 3, 4) is $2\frac{1}{2}$ and the median is $2\frac{1}{2}$. If we include 1,000, the average increases to 202 and the median increases to 3.

   In problem 60, if we exclude the extreme value 182, the average and median of nineteen 2's are both 2. If we include 182, the average increases to 11, and the median remains 2.

   In problem 61, if we exclude the extreme value 45, the average and median of four 10's are both 10. If we include 45, the average increases to 17, and the median remains 10.

   In problem 62, if we exclude the extreme value 0, the average and median of four 100's are both 100. If we include 0, the average decreases to 80, and the median remains 100.

In all of these lists, most of the numbers were close together. Just one very-large or very-small value "pulled" the average far away from the rest of the numbers, but it did not have much of an effect on the median.

So, **the average is more impacted by one extreme value than the median.**

We often say that the average is "more sensitive to extremes" than the median.

In the cases where we have a few very-large or very-small extreme values in our data, the average does not always give us the best picture of the data.

*We often call these extreme values **outliers**. Scientists sometimes choose to ignore outliers when analyzing data. What are some advantages and disadvantages to this choice?*

64. Monster Milkshakes serves $5 \cdot 78 = 390$ people on weekdays (Mon-Fri) and $2 \cdot 120 = 240$ people on weekends (Sat-Sun).

   So, over the 7 days of a week, they serve a total of $390 + 240 = 630$ people. The average daily number of customers served is $\frac{630}{7} =$ **90**.

65. The adult kangaroosters leap a total of $8 \cdot 30 = 240$ feet. The juvenile kangaroosters leap a total of $4 \cdot 21 = 84$ feet. So, all twelve kangaroosters leap a total of $240 + 84 = 324$ feet, and the average leaping distance is $\frac{324}{12} =$ **27 feet**.

66. **Since there are more band members than Beastball players, the average height of all 50 students is closer to the average height of the band members.**

   For example, if the average Beastball player height is 50 inches and the average band member height is 100 inches, then the sum of the student heights is $(10 \cdot 50) + (40 \cdot 100) = 500 + 4,000 = 4,500$ inches. In this case, the average height is $\frac{4,500}{50} = 90$ inches, which is closer to 100 than to 50.

   On the other hand, if the average Beastball player height is 100 inches and the average band member height is 50 inches, then the sum of the student heights is $(10 \cdot 100) + (40 \cdot 50) = 1,000 + 2,000 = 3,000$ inches. In this case, the average height is $\frac{3,000}{50} = 60$ inches, which is closer to 50 than to 100.

67. In the first three rounds, Maggie scored a total of $3 \cdot 8 = 24$ points. In all ten rounds, Maggie scored a total of $10 \cdot 15 = 150$ points. So, in the last 7 rounds, she scored $150 - 24 = 126$ points.

   Therefore, her average number of points per round for the last 7 rounds is $\frac{126}{7} =$ **18**.

68. The original six team members weigh a total of $6 \cdot 15 = 90$ pounds. All nine team members weigh a total of $9 \cdot 12 = 108$ pounds. So, Jake, Ann, and Ungor weigh a total of $108 - 90 = 18$ pounds. Their average weight is therefore $\frac{18}{3} =$ **6 lbs**.

69. The original forty cards in Dave's collection are worth a total of $40 \cdot 6 = 240$ dollars. His 48-card collection is worth a total of $48 \cdot 8 = 384$ dollars. So, his 8 new cards are worth a total of $384 - 240 = 144$ dollars. Their average value is $\frac{144}{8} = \textbf{18 dollars}$.

70. The 15 gold coins are worth $15 \cdot 6 = 90$ dollars. The 33 silver coins are worth $33 \cdot 6 = 198$ dollars. The 57 bronze coins are worth $57 \cdot 6 = 342$ dollars.

All 105 coins are worth a total of $90 + 198 + 342 = 630$ dollars. So, the average value is $\frac{630}{105} = \textbf{6 dollars}$.

— *or* —

To find the average value of all Jack's coins, we find the total value of all the coins and divide by the total number of coins.

Jack's gold coins have an average value of $6. We can replace the value of each gold coin with their average value of $6 without changing the total value of all the gold coins. So, we can assume the value of every gold coin is $6.

Similarly, we can assume every silver coin is worth $6, and that every bronze coin is worth $6 without changing the total value of all the coins.

So, we can assume every coin is worth $6. The average value of any number of $6 coins is **$6**.

71. The 222 tulips have a total height of $222 \cdot 25 = 5{,}550$ cm.

The 333 daffodils have a total height of $333 \cdot 5 = 1{,}665$ cm.

The 555 flowers have a total height of $1{,}665 + 5{,}550 = 7{,}215$ cm.

So, the average height of a flower is $\frac{7{,}215}{555} = \textbf{13 cm}$.

— *or* —

The numbers of flowers are both multiples of 111.

So, we can split these flowers into 111 identical groups. Each group will have $\frac{222}{111} = 2$ tulips and $\frac{333}{111} = 3$ daffodils.

If we replace the height of each tulip with the average height of 25 cm, then we do not change the sum of the tulip heights.

Similarly, if we replace the height of each daffodil with the average height of 5 cm, then we do not change the sum of the daffodil heights.

So, the combined heights of 2 tulips and 3 daffodils is $2 \cdot 25 + 3 \cdot 5 = 50 + 15 = 65$ cm. The average height of each group of five flowers is therefore $\frac{65}{5} = 13$ cm.

Combining identical groups that have the same average does not affect the overall average height. So, the average value of each of the 111 identical groups is the same as the average of all 555 flowers: **13 cm**.

STATISTICS

How Many?   22-23

72. a. Ernie's average score after 2 games is $\frac{0+60}{2} = \frac{60}{2} = \textbf{30}$ points.

b. Ernie's average score after 5 games is $\frac{0 + 4 \cdot 60}{5} = \frac{240}{5} = \textbf{48}$ points.

c. We compare the scores in each game to the 54-point average. The 0-point game is 54 points below the average, and each 60-point game is 6 points above the average.

$$
\begin{array}{ccc}
-54 & +54 & \\
\overline{-54} & \overline{+6 \;\cdots\; +6} & \\
0 & 60 \cdots 60 &
\end{array}
$$

Ernie needs $\frac{54}{6} = 9$ sixty-point games to raise his average to 54 points, so he will play $1 + 9 = \textbf{10}$ games all together.

— *or* —

We write two expressions for the number of points Ernie has scored and solve an equation.

Let g be the number of sixty-point games Ernie played to get an average score of 54.

- Ernie has played 1 zero-point game and g sixty-point games. So, he has scored $0 + 60g = 60g$ points.

- Ernie's average score is 54 after $g + 1$ games, so he has scored at total of $54(g + 1)$ points.

We have two expressions for the total number of points that he scored, so we write an equation:

$$60g = 54(g + 1).$$

We distribute the 54 to get $60g = 54g + 54$. Solving for g, we find $g = 9$.

So, after a total of $9 + 1 = \textbf{10}$ games, Ernie's average is 54 points.

d. We compare the scores in each game to the 59-point average. The 0-point game is 59 points below the average, and each 60-point game is 1 point above the average.

$$
\begin{array}{ccc}
-59 & +59 & \\
\overline{-59} & \overline{+1 \;\cdots\; +1} & \\
0 & 60 \cdots 60 &
\end{array}
$$

Ernie needs 59 sixty-point games to raise his average to 59 points, so he will play $1 + 59 = \textbf{60}$ games all together.

— *or* —

We write two expressions for the number of points Ernie has scored and solve an equation.

Let g be the number of sixty-point games Ernie played to get an average score of 59.

- Ernie has played 1 zero-point game and g sixty-point games. So, he has scored $0 + 60g = 60g$ points.

- Ernie's average score is 59 after $g + 1$ games, so he has scored at total of $59(g + 1)$ points.

We have two expressions for the number of points that he scored, so we write an equation:

$$60g = 59(g + 1).$$

We distribute the 59 to get $60g = 59g + 59$.
Solving for g, we find $g = 59$.

So, after a total of $59 + 1 = 60$ games, Ernie's average is 59 points.

73. Each nickel is 4 cents below the average, and each quarter is 16 cents above the average.

So, the 12 nickels are a total of $12 \cdot 4 = 48$ cents below the average. To balance this, we need enough quarters to equal 48 cents above the average.

$$\overbrace{\underset{5\ \ 5\ \ 5\ \ 5\ \ 5\ \ 5\ \ 5\ \ 5\ \ 5\ \ 5}{-4\ -4\ -4\ -4\ -4\ -4\ -4\ -4\ -4\ -4}}^{-48} \qquad \overbrace{\underset{25\ \cdots\ 25}{+16\ \cdots\ +16}}^{+48}$$

So, the bag contains $\frac{48}{16} = \mathbf{3}$ quarters.

— *or* —

We write and solve an equation. Let q be the number of quarters in the bag.

- The nickels in the bag are worth a total $12 \cdot 5 = 60$ cents, and the quarters are worth a total of $25q$ cents. So, the coins are worth a total of $60 + 25q$ cents.

- There are $q + 12$ coins in the bag with an average value of 9 cents, so the coins are worth a total of $9(q + 12)$ cents.

We have two expressions for the total value of the coins in the bag, so we write an equation:
$$60 + 25q = 9(q + 12).$$

Distributing the 9, we get $60 + 25q = 9q + 108$.
Solving for q, we find $q = 3$.

So, there are **3** quarters in the bag.

74. Each 6 is six above the desired average, and each -2 is two below the desired average. The eight 6's are a total of $8 \cdot 6 = 48$ above the average. To balance this, we need enough -2's to equal 48 below the average.

$$\overbrace{\underset{6\ \ 6\ \ 6\ \ 6\ \ 6\ \ 6\ \ 6}{+6\ +6\ +6\ +6\ +6\ +6\ +6\ +6}}^{+48} \qquad \overbrace{\underset{\text{-2}\ \cdots\ \text{-2}}{-2\ \cdots\ -2}}^{-48}$$

Therefore, we must add $-48 \div (-2) = \mathbf{24}$ negative twos to a list of 8 sixes so that the average of the list is 0.

— *or* —

We write and solve an equation. Let t be the number of negative twos that we add.

- The eight 6's have a sum of $8 \cdot 6 = 48$, and the t negative twos have a sum of -2t. So, the sum of eight 6's and t negative twos is $48 + (-2t)$.

- There are $8 + t$ numbers on the list with an average value of 0, so their sum is 0.

We have two expressions for the sum of the numbers in the list, so we write an equation:
$$48 + (-2t) = 0.$$

Solving for t, we get $t = 24$. Therefore, we must add **24** -2's to eight 6's so that the average of the list is 0.

75. Each 100-point assignment is 8 points above the average, and each 70-point assignment is 22 points below the average. So, the twelve 70-point assignments are a total of $12 \cdot 22 = 264$ points below the average. To balance this, we need enough 100-point assignments to equal 264 points above the average.

$$\overbrace{\underset{70\ \ 70\ \ 70\ \ 70\ \ 70\ \ 70\ \ 70\ \ 70\ \ 70\ \ 70\ \ 70\ \ 70}{-22\ -22\ -22\ -22\ -22\ -22\ -22\ -22\ -22\ -22\ -22\ -22}}^{-264}$$

$$\overbrace{\underset{100\ \cdots\ 100}{+8\ \cdots\ +8}}^{+264}$$

So, Penelope turned in $\frac{264}{8} = 33$ one-hundred-point assignments. Therefore, she turned in a total of $33 + 12 = \mathbf{45}$ assignments.

— *or* —

We write and solve an equation. Let a be the number of assignments that Penelope turned in. Since 12 of these assignments earned 70 points each, $a - 12$ assignments earned 100 points each.

- Penelope's 70-point assignments earned a total of $12 \cdot 70 = 840$ points, and her 100-point assignments earned a total of $100(a - 12)$ points. So, she has earned $840 + 100(a - 12)$ points all together.

- She turned in a assignments that earned an average of 92 points, so she earned a total of $92a$ points.

We have two expressions for the total number of points she earned, so we write an equation:
$$840 + 100(a - 12) = 92a.$$

Distributing the 100, we get $840 + 100a - 1,200 = 92a$.

Solving for a, we find $a = 45$. So, she turned in **45** assignments.

76. We compare the average weight of the flamingoats that Trina did not adopt to the average weight of all the flamingoats.

If we replace the weight of a flamingoat that Trina did not adopt with the average of 3 pounds, then we do not change the sum of the weights of the flamingoats.

Each of the 3-pound flamingoats is 5 pounds below the average weight of all the flamingoats. Trina's 28-pound flamingoat is 20 pounds above the average weight of all flamingoats.

$$\overbrace{\underset{3\ \cdots\ 3}{-5\ \cdots\ -5}}^{-20} \qquad \overbrace{\underset{28}{+20}}^{+20}$$

So, there were $\frac{20}{5} = 4$ flamingoats with an average of weight of 3 pounds and one 28-pound flamingoat. All together, there were $4 + 1 = \mathbf{5}$ flamingoats at the shelter on Tuesday morning.

— *or* —

We write and solve an equation. Let f be the number of flamingoats at the shelter Tuesday morning. There are $f-1$ flamingoats left after Trina adopts one.

- The f flamingoats weigh an average of 8 pounds each, so the total weight of the flamingoats at the shelter on Tuesday morning is $8f$.

- The $f-1$ flamingoats that Trina did not adopt have an average weight of 3 pounds, so they weigh a total of $3(f-1)$ pounds. The flamingoat Trina adopted weighs 28 pounds. So, $3(f-1)+28$ is the total weight of the flamingoats at the shelter on Tuesday morning.

We have two expressions for the total weight of the flamingoats at the shelter on Tuesday morning, so we write an equation:

$$8f=3(f-1)+28.$$

Distributing the 3, we get $8f=3f-3+28$.

Solving for f, we find $f=5$. So, **5** flamingoats were at the shelter on Tuesday morning.

77. We compare each monster's age to the new average.

The average age of all the monsters at the reunion before Betsy and her son arrive is 36. If we replace the age of each of these monsters with 36, we do not change the sum of the ages of these monsters.

Each 36-year-old monster (including Cousin Betsy) is 1 year above the new average age of 35, and Betsy's 11-year-old son is 24 years below the new average age.

$$\begin{array}{ccc} \overbrace{+1 \quad \cdots \quad +1}^{+24} & \overbrace{-24}^{-24} \\ 36 \ \cdots \ 36 & 11 \end{array}$$

So, after Betsy and her son arrive, we have 24 monsters whose average age is 36 and 1 monster who is 11. All together, there are now $24+1=25$ monsters at the reunion.

— *or* —

We write and solve an equation. Let m be the number of monsters there *before* Betsy and her son arrive.

- Betsy's and her son's ages have a sum of $36+11=47$, and the sum of the ages of the other monsters is $36m$. So, the sum of all the monsters' ages is $47+36m$.

- After Betsy and her son arrive, there are $m+2$ monsters at the reunion with an average age of 35, so the sum of the ages of all the monsters is $35(m+2)$.

We have two expressions for the sum of the ages of all the monsters, so we write an equation:

$$47+36m=35(m+2).$$

Distributing the 35, we have $47+36m=35m+70$. Solving for m, we get $m=23$.

So, there are $23+2=$ **25** monsters now at the reunion.

STATISTICS
Mode and Range   24-25

78. The number 20 appears twice in this list, while all other numbers appear once. So, the mode is **20**.

79. The numbers 2 and 6 each appear twice, while the other numbers appear once. So, the modes are **2 and 6**.

80. The number 5 appears three times, while all other numbers appear once or twice. So, the mode is **5**.

81. Each number in this list appears exactly once. This list has **no mode**.

82. The smallest number in the list is -32, and the largest number is 48. Their difference is $48-(-32)=48+32=80$. So, the range of the list is **80**.

83. The least number in the list is 1.2, and the greatest is 4.0. Their difference is $4.0-1.2=2.8$, so the range is **2.8**.

84. Since the mode age is 6, at least two of the monsters are 6 years old.

The range of the ages is 7 years. So, if the two oldest monsters were 6, then the youngest monster would be $6-7=-1$. However, "-1 years old" is not an age!

Therefore, the two youngest monsters must be 6, and the oldest monster is $6+7=$ **13 years old**.

85. The median of the three numbers is 7. Since there is no mode, one of the other numbers is less than 7, and the other number is greater than 7:

$$\underline{\quad}, 7, \underline{\quad}$$

Three numbers whose average is 6 have a sum of $3 \cdot 6 = 18$, so the sum of the other two numbers in the list is $18-7=11$.

There are many pairs of integers whose sum is 11, where one number is less than 7 and one number is greater than 7. However, we get the smallest range when the smallest and largest numbers are as close as possible. (3 and 8) is the closest such pair.

So, the list with the smallest possible range is (3, 7, 8), and its range is $8-3=$ **5**.

86. For there to be a mode, x must be 100, 400, 500, or 600. Of these values, only 400 or 500 could be the median. So, the, median and mode are both either 400 or 500.

If the mean is 400, then the sum of the five numbers is $5 \cdot 400 = 2,000$. In this cas, x is
$$2,000-100-500-600-400= \mathbf{400}.$$

We check that the list (100, 400, 400, 500, 600) has the same mean, median, and mode.

Mean: $\frac{2,000}{5}=400$.   Median: 400.   Mode: 400. ✓

If the mean is 500, then the sum of the five numbers is $5 \cdot 500 = 2,500$. In this case, x is
$$2,500-100-500-600-400=900.$$

However, (100, 400, 500, 600, 900) has no mode. ✗

87. The median of the thirteen numbers is 15, so 15 is the middle number. Because we know the range, we allow for the largest possible number in this list by making the smallest number in the list as big as possible.

Since the list is 13 *different* integers, 9 is the largest possible value of the smallest number, as shown below.

9, 10, 11, 12, 13, 14, 15, __, __, __, __, __, __

Then, since the range is 17, the largest number in this list is $9+17=$ **26**.

Any other list with a range of 17 whose smallest integer is less than 9 has a largest integer that is less than 26.

88. Since the median flower height is 81 centimeters, 81 is the middle number when we write the heights in order. The weights are distinct integers, so 3 flowers are shorter than 81 cm, and 3 that are taller than 81 cm.

$$\underline{\ \ }, \underline{\ \ }, \underline{\ \ }, 81, \underline{\ \ }, \underline{\ \ }, \underline{\ \ }$$

We want the greatest possible range for this list, which we get by making the smallest number as small as possible and the largest number as large as possible.

The heights are all positive integers, so the smallest possible height is 1:

$$1, \underline{\ \ }, \underline{\ \ }, 81, \underline{\ \ }, \underline{\ \ }, \underline{\ \ }$$

Then, since the mean height is 80, the sum of all seven heights is $7 \cdot 80 = 560$. So, to allow the largest number in this list to be as big as possible, we make the rest of the numbers as small as possible.

$$1, 2, 3, 81, 82, 83, \underline{\ \ },$$

In this case, the tallest flower has a height of $560-1-2-3-81-82-83 = 308$ cm.

Therefore, the greatest possible range of heights is $308-1=$ **307** cm.

89. Since the average height of the entire class is 28 inches, the sum of all ten heights is 280 inches.

The average height decreases to $28-6=22$ inches when the tallest monster is removed, so the sum of the remaining nine heights is $9 \cdot 22 = 198$ inches. The tallest monster is therefore $280-198=82$ inches tall.

The average increases to 30 inches when the shortest monster is removed, so the sum of the remaining nine heights is $9 \cdot 30 = 270$ inches. The shortest monster is therefore $280-270=10$ inches tall.

The range of the heights is $82-10=$ **72** inches.

Think about it: Does this range change if the average height of the class is different?

STATISTICS

Stat Squares 26-27

90. The average of the numbers in the middle column is 5, so the sum of these three numbers is $3 \cdot 5 = 15$. Therefore, the empty square in the middle column is $15-7-3=5$.

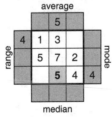

The mode of the bottom row is 4, so the empty square in the bottom row is 4.

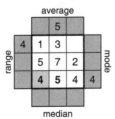

Finally, the range of the numbers in the top row is 4. The smallest digit is 1, so the empty square in the top row is $1+4=5$.

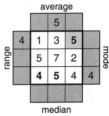

91. The median of the middle column is 1, so we place a 1 in the empty square in the middle column.

The mode of the top row is 3, so the empty square in the top row is 3.

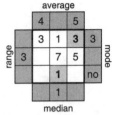

The average of the numbers in the right column is 5, so the sum of these three numbers is $3 \cdot 5 = 15$. Therefore, the empty square in the right column is $15-3-5=7$.

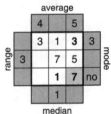

The range of the middle row is 3. Either the 5 is the smallest digit and $5+3=8$ is the largest digit, or the 7 is the largest digit and $7-3=4$ is the smallest digit.

So, the empty square in the middle row is either 8 or 4.

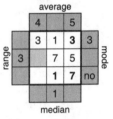

The average of the left column is 4, so the sum of the digits in the left column is 12.

If the empty square in the middle row is 8, then the other empty square in the left column is $12-3-8=1$.

However, another 1 in the bottom row conflicts with the "no" mode clue for the bottom row.

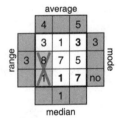

So, the empty square in the middle row is $7-3=4$.

Then the other empty square in the left column is $12-3-4=5$.

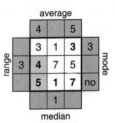

92. The sum of the two empty squares of the middle column is $21-3=18$. There is only one way to get a sum of 18 from two digits: $9+9$. So, both empty squares in the middle column are 9's.

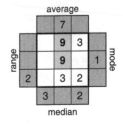

The mode of the middle row is 1, so both empty squares in the middle row are 1's.

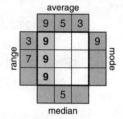

The range of the bottom row is 2, so the empty square in the bottom row is 1 or 4.

The median of the left column is 3, and this column already contains a 1. So, the bottom square of the left column must be 4. Then, we complete the puzzle with a 3 in the upper-left square.

93. The bottom row has a range of 6. If 4 were the smallest number in this row, then the largest would be 4+6 = 10, which is not a digit. So, 8 is the largest number in the bottom row, and the empty square in this row is 8−6 = 2.

The median of the middle column is 5, so 5 must appear in this column. The top row has no mode, so we cannot place another 5 in the top row. Therefore, the 5 in the middle column must be placed in the middle square, as shown.

Then, the mode of the middle row is 2, so the two empty squares in the middle row are 2's.

Finally, the average of the middle column is 6, so the sum of these three numbers is 3·6 = 18. Therefore, the empty square in the middle column is 18−5−4 = 9.

94. The mode of the bottom row is 9, and the range is 5. So, the three digits in the bottom row are (4, 9, 9) in some order.

The average of the middle column is 3, so the sum of these digits is 9. Therefore, we cannot have a 9 in the middle column, and the digit in the middle square of the bottom row is 4.

Then, we use the reasoning discussed in previous puzzles to complete this puzzle as shown.

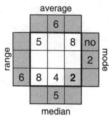

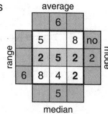

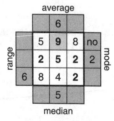

95. The only way to get an average of 9 from three digits is with three 9's, so each square in the left column is 9.

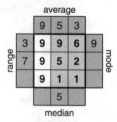

Then, we use the reasoning discussed in previous puzzles to complete this puzzle as shown.

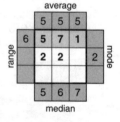

96. In the right column, the only way to get a sum of 3·5 = 15 from three numbers with a median of 7 is 1+7+7. So, the three digits in the right column are (1, 7, 7) in some order.

The mode of the middle row is 2, and the right column contains only 1's and 7's. So, the left and middle squares of the middle row are both 2's.

We use the average and median clues for the left column to determine that the remaining digits in this column are 5 and 8.

Similarly, we use the average and median clues for the middle column to determine that the remaining digits in this column are 6 and 7.

Therefore, in the top row, we have 5 or 8 in the left square, 6 or 7 in the middle square, and 1 or 7 in the right square.

There is only one way to select a choice for each square so that the range of the top row is 6, which is shown to the right.

Then, we complete the puzzle as shown.

97. The median of the right column is 8. So, the two empty squares in the right column are each at least 8.

The range of the middle row is 5. Since the digit in the right square of this row is at least 8, the smallest number in this row is at least $8-5=3$.

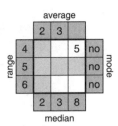

In the left column, there are two ways to get a sum of $2 \cdot 3 = 6$ from three digits with a median of 2:

$$1+2+3 \quad \text{or} \quad 2+2+2.$$

Since every number in the middle row is at least 3, and every number in the left column is 3 or less, the digit in the left square of the middle row is 3.

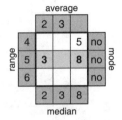

Then, the right square of the middle row is $3+5=8$.

Similarly, the digit in the left square of the bottom row is 2, and the digit in the right square of the bottom row is 8.

The remaining empty square in the left column is 1.

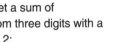

In the middle column, there are three ways to get a sum of $3 \cdot 3 = 9$ from three digits with a median of 3:

$$1+3+5, \quad 2+3+4, \quad \text{or} \quad 3+3+3.$$

Of the three sums above, we can only place (2, 3, 4) as shown so that each row has no mode and the given range.

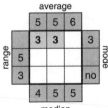

98. In the right column, there are two ways to get a sum of 18 from three numbers with a median of 5:

$$4+5+9 \quad \text{or} \quad 5+5+8.$$

So, the digits in the right column are (4, 5, 9) or (5, 5, 8), in some order.

The mode of the top row is 3. From the sums above, we see that each digit in the right column is at least 4. So, the left and middle squares of the top row are both 3.

We use the average and median clues for the left column to determine that the remaining digits in this column are 4 and 8.

Similarly, we use the average and median clues for the middle column to determine that the remaining digits in this column are 5 and 7.

Therefore, the middle row has 4 or 8 in the left square, and 5 or 7 in the middle square.

Above, we found that the digits in the right column are (4, 5, 9) or (5, 5, 8). So, the right square of the middle row is 4, 5, 8, or 9.

In the middle row, only choosing 4 for the left square and 9 for the right square gives this row a range of 5.

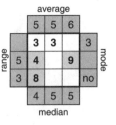

The remaining empty square in the left column is 8.

Previously, we found that the remaining numbers in the middle column are 5 and 7, so the middle square of the bottom row is 5 or 7.

We also previously found that the digits in the right column are (4, 5, 9) or (5, 5, 8). Since there is a 9 in the middle square of the right column, the remaining numbers in the right column are 4 and 5. So, the right square of the bottom row is 4 or 5.

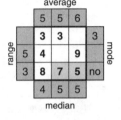

Only choosing 7 for the middle square and 5 for the right square gives the bottom row a range of 5 *and* no mode.

Finally, we fill in the remaining square of each row as shown to complete the puzzle.

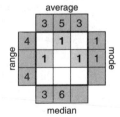

99. We use the mode and range clues for the top row to determine that this row contains two 1's and a 5.

In the middle column, the average tells us that the sum of the three numbers is 15. There are a few ways to get a sum of 15 from three digits with a median of 6, but only one way includes a 1 or a 5: $1+6+8$.

Since the top digit in this row must be a 1 or a 5, the middle column contains 1, 6, and 8 in some order.

So, the top square in the middle column is 1.

The other two squares in the middle column are 6 and 8. The middle row has a mode of 1, so the left and right squares of the middle row are both 1's.

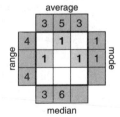

We use the average and median clues for the left column to determine that the two remaining digits in this column are 3 and 5.

Since the missing digits in the top row are 1 and 5, we place the 5 in the top square of the left column and the 3 in the bottom square.

Then, we complete the top row with a 1.

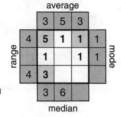

The remaining empty square in the right column is therefore $9-1-1=7$.

Finally, we can only place the 6 and 8 in the middle column as shown so that the range of the bottom row is 4.

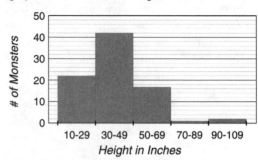

STATISTICS
Data Display
28-29

100. The height of a bar is given by the number of monsters, with taller bars indicating more monsters whose heights are in that range. The tallest bar is the "40-49" bar. So, **40-49** inches is the range that contains the greatest number of monsters.

To help us answer the questions below, we make a list of the number of monsters in each range:

10-19	20-29	30-39	40-49	50-59	60-69	70-79	80-89	90-99
7	15	19	23	13	4	1	0	2

101. One monster is 70-79 inches tall and 2 monsters are 90-99 inches tall. So, **3** of these monsters are at least 70 inches tall.

102. There are $7+15+19+23+13+4+1+0+2 = $ **84** heights displayed in the histogram.

103. The median height of the 84 monsters is the number exactly between the two middle heights. If we order all of these monsters from shortest to tallest, the two middle monsters will be in the 42nd and 43rd positions.

$7+15+19=41$ monsters are 10-39 inches tall, and 23 monsters are 40-49 inches tall. Therefore, the monsters in the 42nd and 43rd positions are both 40-49 inches tall. So, the number exactly between these two heights is also in the range 40-49.

43 is the only choice between 40 and 49. So, among these eight numbers, only **43 inches** *could* be the median height of the monsters.

15 in   22 in   29 in   36 in   (43 in)   50 in   57 in   64 in

104. We find the height of each bar in the new histogram by adding the number of monsters in each range.

Our original histogram only goes up to 99 inches, so we may assume that no 10-year-olds at Beast Academy are taller than 99 inches.

10-29	30-49	50-69	70-89	90-109
7+15	19+23	13+4	1+0	2+0
22	42	17	1	2

We graph this data on the histogram as shown.

[Histogram: # of Monsters vs Height in Inches, with bars 10-29: 22, 30-49: 42, 50-69: 17, 70-89: 1, 90-109: 2]

105. We compute each student's average score.

James: $\frac{100+0+100+100+40+40}{6} = \frac{380}{6} = 63\frac{1}{3}$

Alan: $\frac{60+80+60+80+70+80}{6} = \frac{430}{6} = 71\frac{2}{3}$

Mika: $\frac{10+60+70+90+100+90}{6} = \frac{420}{6} = 70$

Pierre: $\frac{90+90+80+70+60+50}{6} = \frac{440}{6} = 73\frac{1}{3}$

Pierre has the highest average score.

You may have noticed that we can compare these four average without computing each quotient:

$$\frac{440}{6} > \frac{430}{6} > \frac{420}{6} > \frac{380}{6}.$$

106. We list each student's scores in order from least to greatest to determine each student's median score.

James: 0, 40, 40, 100, 100, 100   Median: $\frac{40+100}{2} = 70$

Alan: 60, 60, 70, 80, 80, 80   Median: $\frac{70+80}{2} = 75$

Mika: 10, 60, 70, 90, 90, 100   Median: $\frac{70+90}{2} = 80$

Pierre: 50, 60, 70, 80, 90, 90   Median: $\frac{70+80}{2} = 75$

Mika has the highest median score.

107. We list each student's mode score.

James: 100   Alan: 80   Mika: 90   Pierre: 90

James has the highest mode score.

108. We list the range of each student's scores.

James: $100-0=100$ Alan: $80-60=20$

Mika: $100-10=90$ Pierre: $90-50=40$

Alan has the smallest range of scores.

109. We look at the shape of the graph of each student's scores:

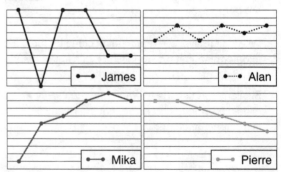

James' scores are sometimes high and sometimes low.
Alan's scores do not change very much.
Pierre's scores have decreased over the six months.

Only **Mika** started out with lower scores and has increased his score nearly every month after.

110. We could make an argument for any of these four students.

James: James has the highest mode score, but his scores also change a lot.

Alan: Alan's scores have the smallest range, so his scores do not change very much. But, those scores are not very high.

Mika: Mika's scores have improved the most over the last six months, but he has the lowest average, median, and mode of all four students.

Pierre: Pierre has the highest average score, but his scores have been declining over the last few months.

111. If we make a list of all student scores over all six months, the list of 24 scores has four modes: 60, 80, 90, and 100. Each of those scores appears four times.

If Mr. Erikson finds that the only mode of three students' scores is 80, then the student Mr. Erikson left out received at least one 60, one 90, and one 100. Only Mika received all three of these scores, so Mr. Erikson did not include **Mika**'s scores.

112. We first compute the median score for each month.

Sept: $\frac{60+90}{2} = \frac{150}{2} = 75$ Oct: $\frac{60+80}{2} = \frac{140}{2} = 70$

Nov: $\frac{70+80}{2} = \frac{150}{2} = 75$ Dec: $\frac{80+90}{2} = \frac{170}{2} = 85$

Jan: $\frac{60+70}{2} = \frac{130}{2} = 65$ Feb: $\frac{50+80}{2} = \frac{130}{2} = 65$

To graph this data, we place a dot above each month's label at the height of that month's median score, then connect consecutive dots with a line segment.

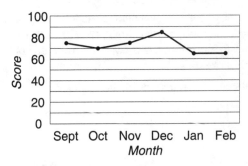

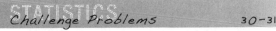

STATISTICS
Challenge Problems   30-31

113. We consider the way the numbers balance around the median. Depending on whether we have an even or odd number of integers, s is either the middle number of the list or the number exactly between the two middle numbers. Because our list contains consecutive integers, for every number in the list that is greater than s, there is one number that is equally less than than s.

So, these numbers balance perfectly around the median, and the average of the list is equal to the median, s.

Examples:
(2, 3, 4, 5, 6, 7, 8) has both a median and average of 5.
(10, 11, 12, 13) has both a median and average of $11\frac{1}{2}$.

114. The six numbers we can create by arranging the digits 1, 2, and 3 are 123, 132, 213, 231, 312, and 321.

The average of these six numbers is

$\frac{123+132+213+231+312+321}{6} = \frac{1,332}{6} = \mathbf{222}.$

— *or* —

In the six arrangements, the digits 1, 2, and 3 each appear in the hundreds, tens, and ones places twice. The average of (1, 1, 2, 2, 3, 3) is 2. So the hundreds, tens, and ones digits of the average are all 2's: **222**.

115. An average of 3 bridges touch each island, and there are 6 islands. So, there are $3 \cdot 6 = 18$ bridge ends on all the islands.

Each bridge has two ends, so there are $\frac{18}{2} = \mathbf{9}$ bridges that connect the six islands.

To check, we draw an example with circles representing islands and curved lines representing bridges.

There are 6 islands and 9 bridges, and the average number of bridges that touch an island is

$\frac{3+3+4+5+1+2}{6} = \frac{18}{6} = 3.$ ✓

116. To compute the average of these three expressions, we add them together and then divide by 3. Their sum is

$$(3x+y)+(3y+z)+(3z+x) = 4x+4y+4z.$$

We can factor a 4 out of each term with a variable to get

$$4x+4y+4z = 4(x+y+z).$$

Since the average of x, y, and z is 6, we know the sum $x+y+z$ is 18. So, the sum of the three expressions is equal to $4(x+y+z) = 4(18) = 72$.

The average of the three expressions is therefore

$$\frac{4x+4y+4z}{3} = \frac{72}{3} = \mathbf{24}.$$

117. The difference between the largest and smallest numbers in this list of consecutive integers is 31. So, there are 32 numbers on this list.

In problem 113, we found that the average and median of any list of consecutive integers are equal. So, $10\frac{1}{2}$ is the number halfway between the two middle numbers of this list: 10 and 11.

Therefore, this list of 32 consecutive integers contains 15 numbers greater than 11 and 15 numbers less than 10. The smallest number is 15 less than ten: $10-15 = \mathbf{-5}$.

118. We first order the known values:

$$12, 17, 21, 30.$$

When we add the fifth value, n, we have three possibilities for the median of the five-number group.

- If n is less than 17, then the median is 17:

$$12, n, 17, 21, 30 \quad \textbf{\textit{or}} \quad n, 12, 17, 21, 30.$$

Since we want the median to equal the average, we write the following equation:

$$\frac{12+17+21+30+n}{5}=17.$$

Solving for n, we get $n=5$. The list (5, 12, 17, 21, 30) has median 17 and average 17. ✓

- If n is at least 17 but not more than 21, then n is the median:

$$12, 17, n, 21, 30.$$

Since we want the median to equal the average, we write the following equation:

$$\frac{12+17+21+30+n}{5}=n.$$

Multiplying both sides of the equation by 5 gives us $12+17+21+30+n=5n$.

Solving for n, we get $n=20$. The list (12, 17, 20, 21, 30) has median 20 and average 20. ✓

- If n is greater than 21, then the median is 21:

$$12, 17, 21, n, 30 \quad \textbf{\textit{or}} \quad 12, 17, 21, 30, n.$$

Since we want the median to equal the average, we write the following equation:

$$\frac{12+17+21+30+n}{5}=21.$$

Solving for n, we get $n=25$. The list (12, 17, 21, 25, 30) has median 21 and average 21. ✓

So, the median equals the average of the list (17, 12, 21, 30, n) when n is **5, 20, or 25**.

119. The number 5 appears twice in this list, while each other number appears once. So, 5 must be the mode of the six-number list.

We also note that when a is the smallest number on the list, the median is 6: (a, 5, 5, 7, 8, 13). There is no way to get a smaller median, so the median of this six-number list is always greater than 5.

There are only two groups of 3 consecutive numbers that include 5 and at least one number greater than 5: (4, 5, 6) and (5, 6, 7).

(4, 5, 6): The median is greater than 5, so it can only be 6. The mode is 5. This leaves 4 as the mean, and we have

$$\frac{5+5+7+8+13+a}{6}=4.$$

Solving for a, we get $a=$ -14. The list (-14, 5, 5, 7, 8, 13) has mean 4, mode 5, and median 6. ✓

(5, 6, 7): The mode is 5. The median is more than 5, so it could be 6 or 7.

If the median is 6, then the mean is 7 and we have

$$\frac{5+5+7+8+13+a}{6}=7.$$

Solving for a, we get $a=4$. The list (4, 5, 5, 7, 8, 13) has mode 5, median 6, and mean 7. ✓

If the median is 7, then the mean is 6 and we have

$$\frac{5+5+7+8+13+a}{6}=6.$$

Solving for a, we get $a=$ -2. The list (-2, 5, 5, 7, 8, 13) has mode 5, median 6, and mean 6. ✗

You may have instead noticed that for the median of (a, 5, 5, 7, 8, 13) to be 7, we must have $a=7$. However, (5, 5, 7, 7, 8, 13) has two modes. ✗

Therefore, **-14 and 4** are the only values of a for which the mean, median, and mode of the list are consecutive integers.

120. When Paulie's list is written in order, the middle number is 20. Since the mode is 21, the list has two or three 21's:

$$\_\_, \_\_, \_\_, 20, 21, 21, \_\_.$$

Also, the average of Paulie's list is 18, so the sum of the seven numbers on her list is $7 \cdot 18=126$.

- If the list has exactly two 21's, then every other number in the list appears exactly once. We fill in each of the four remaining spots with the smallest possibilities so that the list has a range of 7:

$$\underline{15}, \underline{16}, \underline{17}, 20, 21, 21, \underline{22}.$$

The sum of these numbers is 132, which is greater than the sum of Paulie's list. Any list with larger numbers will give an even greater sum. So, 21 must appear more than two times.

- If the list has three 21's, then 21 is the largest number in the list and $21-7=14$ is the smallest:

$$14, \_\_, \_\_, 20, 21, 21, 21.$$

The sum of the known numbers is 97, so the sum of the two remaining numbers is $126-97=29$. The only way to make a sum of 29 with two numbers from 14 to 20 is $14+15$, so these are the two remaining numbers.

Therefore, Paulie's list is **14, 14, 15, 20, 21, 21, 21**. ✓

1. The factors of 18 are **1, 2, 3, 6, 9,** and **18**.

The factors of 48 are **1, 2, 3, 4, 6, 8, 12, 16, 24,** and **48**.

The factors of 37 are **1** and **37**.

The factors of 54 are **1, 2, 3, 6, 9, 18, 27,** and **54**.

The factors of 49 are **1, 7,** and **49**.

The factors of 225 are **1, 3, 5, 9, 15, 25, 45, 75,** and **225**.

2. We list the factors of 140:

1, 2, 4, 5, 7, 10, 14, 20, 28, 35, 70, 140.

Of these factors, only 2, 5, and 7 are prime. So, **7** is the largest prime factor of 140.

3. We list the factors of each number.

<u>80</u>: 1, 2, 4, 5, 8, 10, 16, 20, 40, 80.
<u>56</u>: 1, 2, 4, 7, 8, 14, 28, 56.

The factors of 80 that are also factors of 56 are 1, 2, 4, and 8. So, **4** factors of 80 are also factors of 56.

For each problem below, you may have created a different factor tree to arrive at the same prime factorization.

4. We make a factor tree for 40, circling factors that are prime:

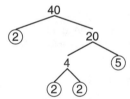

The prime factorization of 40 is $2^3 \cdot 5$.

5. We make a factor tree for 105:

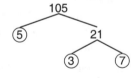

The prime factorization of 105 is $3 \cdot 5 \cdot 7$.

6. We make a factor tree for 66:

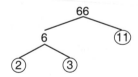

The prime factorization of 66 is $2 \cdot 3 \cdot 11$.

7. Since 43 is prime, its prime factorization is **43**.

8. We make a factor tree for 128:

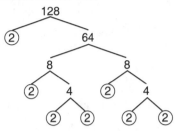

The prime factorization of 128 is 2^7.

9. We make a factor tree for 222:

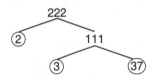

The prime factorization of 222 is $2 \cdot 3 \cdot 37$.

10. We make a factor tree for 364:

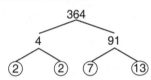

The prime factorization of 364 is $2^2 \cdot 7 \cdot 13$.

11. We make a factor tree for 475:

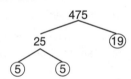

The prime factorization of 475 is $5^2 \cdot 19$.

12. We work backwards from the largest two-digit number, 99.

Since $99 = 3 \cdot 33$, it is composite.

Since $98 = 2 \cdot 49$, it is composite.

97 is not divisible by the primes 2, 3, 5, or 7. The next-smallest prime is 11. Since $11 \cdot 11 = 121$, we would have to multiply 11 by an integer that is less than 11 to get 97. However, we have already tested every prime less than 11. Therefore, 97 is not divisible by 11. By the same reasoning, 97 is not divisible by any other primes less than 97.

Therefore, 97 is prime. So, **97** is the largest two-digit prime number.

Review the divisibility tests used above in the Division chapter of Beast Academy Guide 4B.

13. A number that has 7 as a prime factor is a multiple of 7. So, we count the multiples of 7 between 200 and 300.

203 = 7 · 29 is the smallest multiple of 7 greater than 200.
294 = 7 · 42 is the greatest multiple of 7 less than 300.

So, the number of multiples of 7 between 200 and 300 is the same as the number of integers from 29 to 42:

$$29, 30, 31, ..., 40, 41, 42.$$

Subtracting 28 from each integer in this list, we have

$$1, 2, 3, ..., 12, 13, 14.$$

There are 14 numbers in this list. So, there are **14** multiples of 7 between 200 and 300.

14. The factors of 72 are 1, 2, 3, 4, 6, 8, 9, 12, 18, 24, 36, 72.

Of these factors, only 6, 12, 18, 24, 36, and 72 are multiples of 6. So, **6** factors of 72 are also multiples of 6.

— *or* —

Since 72 = 6 · 12, the product of 6 and any factor of 12 is a factor of 72, as well as a multiple of 6.

The factors of 12 are 1, 2, 3, 4, 6, and 12. Therefore, the factors of 72 that are also multiples of 6 are

$$1 \cdot 6 = 6, \qquad 2 \cdot 6 = 12, \qquad 3 \cdot 6 = 18,$$
$$4 \cdot 6 = 24, \qquad 6 \cdot 6 = 36, \qquad 12 \cdot 6 = 72.$$

All together, **6** factors of 72 are also multiples of 6.

15. Since the rectangle has integer side lengths and area 126 square units, each factor pair of 126 represents a possibility for the rectangle's dimensions.

For example, 2 · 63 = 126, so the rectangle could have length 2 and width 63, or length 63 and width 2. However, switching the length and width does not change the perimeter, so each factor pair of 126 gives one possible perimeter.

There are six factor pairs of 126:

$$(1 \cdot 126), (2 \cdot 63), (3 \cdot 42), (6 \cdot 21), (7 \cdot 18), \text{ and } (9 \cdot 14).$$

Each factor pair above gives a rectangle with a distinct perimeter. So, there are **6** different perimeters that are possible.

16. The smallest factors of $2,772 = 2^2 \cdot 3^2 \cdot 7 \cdot 11$ are, in order, 1, 2, 3, 4, 6, 7, 9, and 12.

The factors 1, 2, 3, $4 = 2^2$, $6 = 2 \cdot 3$, and 7 are all factors of $4,200 = 2^3 \cdot 3 \cdot 5^2 \cdot 7$.

Since 4,200 has only one 3 in its prime factorization, $9 = 3^2$ is not a factor of 4,200. So, the smallest integer that is a factor of 2,772 but not of 4,200 is **9**.

— *or* —

We look for powers of primes that are factors of 2,772, but not of 4,200.

$2,772 = 2^2 \cdot 3^2 \cdot 7 \cdot 11$
$4,200 = 2^3 \cdot 3 \cdot 5^2 \cdot 7$

The prime factorization of 4,200 has more 2's than 2,772, and as many 7's as 2,772, but has fewer 3's and no 11's.

So, any factor of 2,772 that is *not* a factor of 4,200 must have two 3's, one 11, or both in its prime factorization.

The smallest number with two 3's, one 11, or both in its prime factorization is $3^2 = $ **9**.

17. We begin by writing the prime factorization of 54:

$$54 = 2 \cdot 3^3.$$

So, any number that is divisible by 54 must have at least one 2 and three 3's in its prime factorization. We consider each of the given numbers.

- $675 = 3^3 \cdot 5^2$:
  675 does not have any 2's in its prime factorization, so 675 is not divisible by 54.

- $882 = 2 \cdot 3^2 \cdot 7^2$:
  882 has only two 3's in its prime factorization, so 882 is not divisible by 54.

- $1,782 = 2 \cdot 3^4 \cdot 11$:
  1,782 has at least one 2 and three 3's in its prime factorization, so 1,782 is divisible by 54.
  $1,782 = 2 \cdot 3^4 \cdot 11 = (2 \cdot 3^3) \cdot 3 \cdot 11 = (54) \cdot 3 \cdot 11$.

- $2,160 = 2^4 \cdot 3^3 \cdot 5$:
  2,160 has at least one 2 and three 3's in its prime factorization, so 2,160 is divisible by 54.
  $2,160 = 2^4 \cdot 3^3 \cdot 5 = (2 \cdot 3^3) \cdot 2^3 \cdot 5 = (54) \cdot 2^3 \cdot 5$.

We circle the numbers that are divisible by 54:

675	882	**1,782**	**2,160**
$3^3 \cdot 5^2$	$2 \cdot 3^2 \cdot 7^2$	$2 \cdot 3^4 \cdot 11$	$2^4 \cdot 3^3 \cdot 5$

18. The prime factorization of 308 is $2^2 \cdot 7 \cdot 11$.

Any number that is divisible by 308 must have at least two 2's, one 7, and one 11 in its prime factorization. We consider each of the given numbers.

- $3,360 = 2^5 \cdot 3 \cdot 5 \cdot 7$:
  3,360 does not have any 11's in its prime factorization, so 3,360 is not divisible by 308.

- $4,312 = 2^3 \cdot 7^2 \cdot 11$:
  4,312 has at least two 2's, one 7, and one 11 in its prime factorization, so 4,312 is divisible by 308.
  $4,312 = 2^3 \cdot 7^2 \cdot 11 = (2^2 \cdot 7 \cdot 11) \cdot 2 \cdot 7 = (308) \cdot 2 \cdot 7$.

- $6,468 = 2^2 \cdot 3 \cdot 7^2 \cdot 11$:
  6,468 has at least two 2's, one 7, and one 11 in its prime factorization, so 6,468 is divisible by 308.
  $6,468 = 2^2 \cdot 3 \cdot 7^2 \cdot 11 = (2^2 \cdot 7 \cdot 11) \cdot 3 \cdot 7 = (308) \cdot 3 \cdot 7$.

- $9,702 = 2 \cdot 3^2 \cdot 7^2 \cdot 11$:
  9,702 does not have two 2's in its prime factorization, so 9,702 is not divisible by 308.

We circle the numbers that are divisible by 308:

3,360	**4,312**	**6,468**	9,702
$2^5 \cdot 3 \cdot 5 \cdot 7$	$2^3 \cdot 7^2 \cdot 11$	$2^2 \cdot 3 \cdot 7^2 \cdot 11$	$2 \cdot 3^2 \cdot 7^2 \cdot 11$

19. We consider the prime factorization of each number.

- $27 = 3^3$:
  4,095 does not have three 3's in its prime factorization. So, 27 is not a factor of 4,095.

- $91 = 7 \cdot 13$:
  4,095 has a 7 and a 13 in its prime factorization. So, 91 is a factor of 4,095.

- $105 = 3 \cdot 5 \cdot 7$:

  4,095 has a 3, a 5, and a 7 in its prime factorization. So, 105 is a factor of 4,095.

- $225 = 3^2 \cdot 5^2$:

  4,095 does not have two 5's in its prime factorization. So, 225 is not a factor of 4,095.

- $315 = 3^2 \cdot 5 \cdot 7$:

  4,095 has two 3's, a 5, and a 7 in its prime factorization. So, 315 is a factor of 4,095.

We circle the numbers that are factors of 4,095:

27   91   105   225   315

20. We consider the prime factorization of each number.

- $32 = 2^5$:

  5,472 has five 2's in its prime factorization. So, 32 is a factor of 5,472.

- $48 = 2^4 \cdot 3$:

  5,472 has at least four 2's and a 3 in its prime factorization. So, 48 is a factor of 5,472.

- $108 = 2^2 \cdot 3^3$:

  5,472 does not have three 3's in its prime factorization. So, 108 is not a factor of 5,472.

- $119 = 7 \cdot 17$:

  5,472 does not have a 7 or a 17 in its prime factorization. So, 119 is not a factor of 5,472.

- $171 = 3^2 \cdot 19$:

  5,472 has two 3's and a 19 in its prime factorization. So, 171 is a factor of 5,472.

We circle the numbers that are factors of 5,472:

32   48   108   119   171

21. The prime factorization of 75 is $3 \cdot 5^2$.

We can use prime factorization to write 7,425 as the product of 75 and another integer:

$$7,425 = 3^3 \cdot 5^2 \cdot 11 = (3 \cdot 5^2) \cdot (3^2 \cdot 11) = 75 \cdot 99.$$

Since $7,425 = 75 \cdot 99$, we have $7,425 \div 75 = \mathbf{99}$.

22. The prime factorization of 196 is $2^2 \cdot 7^2$. We write 49,392 as the product of 196 and another integer:

$$49,392 = 2^4 \cdot 3^2 \cdot 7^3 = (2^2 \cdot 7^2) \cdot (2^2 \cdot 3^2 \cdot 7) = 196 \cdot 252.$$

Since $49,392 = 196 \cdot 252$, we have $49,392 \div 196 = \mathbf{252}$.

23. The prime factorization of 135 is $3^3 \cdot 5$. We write 3,780 as the product of 135 and another integer:

$$3,780 = 2^2 \cdot 3^3 \cdot 5 \cdot 7 = (3^3 \cdot 5) \cdot (2^2 \cdot 7) = 135 \cdot 28.$$

So, we can multiply 135 by **28** to get 3,780.

24. The prime factorization of 504 is $2^3 \cdot 3^2 \cdot 7$.

To find the largest odd factor of 504, we look for the largest factor that is not divisible by 2. We can compute this by multiplying all of the odd prime factors in the prime factorization of 504.

So, the largest odd factor of 504 is $3^2 \cdot 7 = 63$.

To find the quotient of $504 \div 63$, we note that $504 = 2^3 \cdot (3^2 \cdot 7) = 2^3 \cdot (63)$. Therefore, $504 \div 63 = 2^3 = \mathbf{8}$.

25. The larger the power of 6 that Myrtle divides 64,800 by, the smaller the quotient will be. So, to find the smallest integer quotient, we look for the largest power of 6 that 64,800 is divisible by.

Since $6 = 2 \cdot 3$, we pair 2's and 3's in the prime factorization of 64,800 to make as many factors of 6 as possible.

$$64,800 = 2^5 \cdot 3^4 \cdot 5^2$$
$$= (2 \cdot 2 \cdot 2 \cdot 2 \cdot 2) \cdot (3 \cdot 3 \cdot 3 \cdot 3) \cdot (5 \cdot 5)$$
$$= (2 \cdot 3) \cdot (2 \cdot 3) \cdot (2 \cdot 3) \cdot (2 \cdot 3) \cdot (2 \cdot 5 \cdot 5)$$
$$= 6 \cdot 6 \cdot 6 \cdot 6 \cdot 50$$
$$= 6^4 \cdot 50.$$

So, 6^4 is the largest power of 6 Myrtle can divide 64,800 by to get an integer quotient. Therefore, the smallest integer quotient she can get by dividing 64,800 by a power of 6 is $64,800 \div 6^4 = \mathbf{50}$.

FACTORS & MULTIPLES
Perfect Squares   38-39

26. The prime factorization of each square is shown below.

$81 = 9 \cdot 9 = 3^2 \cdot 3^2 = \mathbf{3^4}$.

$121 = 11 \cdot 11 = \mathbf{11^2}$.

$1,600 = 40 \cdot 40 = (2^3 \cdot 5) \cdot (2^3 \cdot 5) = \mathbf{2^6 \cdot 5^2}$.

$3,600 = 60 \cdot 60 = (2^2 \cdot 3 \cdot 5) \cdot (2^2 \cdot 3 \cdot 5) = \mathbf{2^4 \cdot 3^2 \cdot 5^2}$.

27. We consider each prime factorization.

For $3^2 \cdot 11^2$, we have

$$3^2 \cdot 11^2 = (3 \cdot 3) \cdot (11 \cdot 11)$$
$$= (3 \cdot 11) \cdot (3 \cdot 11)$$
$$= 33 \cdot 33$$
$$= \mathbf{33^2}.$$

For 2^8, we have

$$2^8 = 2 \cdot 2 \cdot 2 \cdot 2 \cdot 2 \cdot 2 \cdot 2 \cdot 2$$
$$= (2 \cdot 2 \cdot 2 \cdot 2) \cdot (2 \cdot 2 \cdot 2 \cdot 2)$$
$$= 16 \cdot 16$$
$$= \mathbf{16^2}.$$

For $2^4 \cdot 3^2 \cdot 7^2$, we have

$$2^4 \cdot 3^2 \cdot 7^2 = (2 \cdot 2 \cdot 2 \cdot 2) \cdot (3 \cdot 3) \cdot (7 \cdot 7)$$
$$= (2 \cdot 2 \cdot 3 \cdot 7) \cdot (2 \cdot 2 \cdot 3 \cdot 7)$$
$$= 84 \cdot 84$$
$$= \mathbf{84^2}.$$

28. If a number can be written as the product of two copies of the same integer, then it is a perfect square. So, we try to write the prime factorization of each number as the product of two identical groups of prime factors.

- $7,776 = 2^5 \cdot 3^5$:

  $2^5 \cdot 3^5 = (2^2 \cdot 3^2) \cdot (2^2 \cdot 3^2) \cdot 2 \cdot 3$. There is no way to split the last 2 and 3 so that we have two identical groups of prime factors. So, 7,776 is not a perfect square.

- $3,136 = 2^6 \cdot 7^2$:

  $2^6 \cdot 7^2 = (2^3 \cdot 7) \cdot (2^3 \cdot 7) = 56 \cdot 56 = 56^2$.

  So, 3,136 is a perfect square.

- $81,796 = 2^2 \cdot 11^2 \cdot 13^2$:
  $2^2 \cdot 11^2 \cdot 13^2 = (2 \cdot 11 \cdot 13) \cdot (2 \cdot 11 \cdot 13) = 286 \cdot 286 = 286^2$.
  So, 81,796 is a perfect square.

- $444,771 = 3^4 \cdot 17^2 \cdot 19$:
  $3^4 \cdot 17^2 \cdot 19 = (3^2 \cdot 17) \cdot (3^2 \cdot 17) \cdot 19$. There is no way to group the 19 so that we have two identical groups of prime factors. So, 444,771 is not a perfect square.

We circle the numbers that are perfect squares:

7,776   ⬭3,136⬭   ⬭81,796⬭   444,771
$2^5 \cdot 3^5$ $2^6 \cdot 7^2$ $2^2 \cdot 11^2 \cdot 13^2$ $3^4 \cdot 17^2 \cdot 19$

29. If a number's prime factors can be split into two identical groups, then it is a perfect square. So, we try to write each expression as the product of two identical groups.

- $x^5 \cdot y^5 = (x^2 \cdot y^2) \cdot (x^2 \cdot y^2) \cdot x \cdot y$. There is no way to group the last x and y so that we have the product of two identical groups. So, $x^5 \cdot y^5$ is not a perfect square.

- $x^{81} = x^{40} \cdot x^{40} \cdot x$. There is no way to group the last x so that we have the product of two identical groups. So, x^{81} is not a perfect square.

- $x^2 \cdot y^3 = (x \cdot y) \cdot (x \cdot y) \cdot y$. There is no way to group the last y so that we have the product of two identical groups. So, $x^2 \cdot y^3$ is not a perfect square.

- $y^{12} = y^6 \cdot y^6 = (y^6)^2$. Since we can write y^{12} as the product of two identical groups, it is a perfect square.

- $x \cdot y$ is the product of two different prime factors, so it is not a perfect square.

- $x^4 \cdot y^{10} = (x^2 \cdot y^5) \cdot (x^2 \cdot y^5) = (x^2 \cdot y^5)^2$. Since we can write $x^4 \cdot y^{10}$ as the product of two identical groups, it is a perfect square.

We circle the expressions that are perfect squares:

$x^5 \cdot y^5$ x^{81} $x^2 \cdot y^3$ ⬭y^{12}⬭ $x \cdot y$ ⬭$x^4 \cdot y^{10}$⬭

30. We first consider Grogg's statement. If a number is a perfect square, then it can be written as the product of two identical groups of prime factors.

So, the number of copies of any prime in the prime factorization of a perfect square is double the number of copies of that prime in each identical group.

For example, the perfect square 144 is the product of two 12's. So, the prime factorization of 144 has double the number of 2's and 3's as the prime factorization of 12:

$$144 = 12 \cdot 12 = (2^2 \cdot 3) \cdot (2^2 \cdot 3) = 2^4 \cdot 3^2.$$

Doubling the number of primes in a prime factorization gives an even number of each prime factor. So, every prime will have an even exponent. Therefore, Grogg's statement is correct.

We next consider Lizzie's statement. If the prime factorization of a number includes only even exponents,

then the number of copies of each prime can be spit into two identical groups.

For example, $2^8 = 2^4 \cdot 2^4$, and $5^6 = 5^3 \cdot 5^3$. So, we have

$$\begin{aligned} 2^8 \cdot 5^6 &= (2^4 \cdot 2^4) \cdot (5^3 \cdot 5^3) \\ &= (2^4 \cdot 5^3) \cdot (2^4 \cdot 5^3) \\ &= (2^4 \cdot 5^3)^2. \end{aligned}$$

We can use the same strategy to write *any* number whose prime factorization includes only even exponents as the product of two identical groups.

So, **both Lizzie's and Grogg's statements are correct**.

31. Since 9 is not prime, we cannot immediately tell if 9^3 is a perfect square. So, we begin by writing 9^3 as the product of prime factors. Since $9 = 3 \cdot 3$, we have

$$\begin{aligned} 9^3 &= (3 \cdot 3)^3 \\ &= (3 \cdot 3) \cdot (3 \cdot 3) \cdot (3 \cdot 3) \\ &= (3 \cdot 3 \cdot 3) \cdot (3 \cdot 3 \cdot 3) \\ &= 27 \cdot 27 \\ &= 27^2. \end{aligned}$$

So, **9^3 is a perfect square. We have $9^3 = 27^2$.**

32. The prime factorization of 180 is $2^2 \cdot 3^2 \cdot 5$.

In order for $180n$ to be a perfect square, the exponent of each prime in its prime factorization must be even. Since 5 is the only prime in the prime factorization of 180 without an even exponent, multiplying 180 by 5 gives a perfect square:

$$\begin{aligned} 180 \cdot 5 &= (2^2 \cdot 3^2 \cdot 5) \cdot 5 \\ &= 2^2 \cdot 3^2 \cdot 5^2 \\ &= (2 \cdot 3 \cdot 5) \cdot (2 \cdot 3 \cdot 5) \\ &= 30 \cdot 30 \\ &= 30^2. \end{aligned}$$

Multiplying 180 by any positive integer less than 5 will not give a perfect square.

So, **5** is the smallest positive value of n that makes $180n$ a perfect square.

33. In order for the product of 6,174 and an integer to be a perfect square, the exponent of every prime in the product's prime factorization must be even.

The prime factorization of 6,174 is $2 \cdot 3^2 \cdot 7^3$. The primes 2 and 7 each have odd exponents (we can write 2 as 2^1). So, multiplying 6,174 by $2 \cdot 7$ gives a perfect square:

$$\begin{aligned} 6,174 \cdot 2 \cdot 7 &= (2 \cdot 3^2 \cdot 7^3) \cdot 2 \cdot 7 \\ &= 2^2 \cdot 3^2 \cdot 7^4 \\ &= (2 \cdot 3 \cdot 7^2) \cdot (2 \cdot 3 \cdot 7^2) \\ &= 294 \cdot 294 \\ &= 294^2. \end{aligned}$$

Multiplying 6,174 by any integer less than $2 \cdot 7$ would result in a product that is not a perfect square. So, $2 \cdot 7 = $ **14** is the smallest positive integer we can multiply 6,174 by to make a perfect square.

34. The exponent of every prime in a perfect square's prime factorization is even. So, to find the *largest* perfect square factor of 23,520, we use the largest even exponent possible for each prime in the prime factorization of 23,520.

We are given that $23{,}520 = 2^5 \cdot 3 \cdot 5 \cdot 7^2$.

There are five 2's in $2^5 \cdot 3 \cdot 5 \cdot 7^2$, so the greatest number of 2's we can use is four.
There is only one 3, so we cannot use any 3's.
There is only one 5, so we cannot use any 5's.
There are two 7's, so we can use two 7's.

So, the largest perfect square factor of 23,520 is $2^4 \cdot 7^2 = 16 \cdot 49 = \textbf{784}$. We note that $784 = 28^2$.

FACTORS & MULTIPLES
GCF
40-41

35. We list the factors of 12 and 15.

12: 1, 2, 3, 4, 6, 12
15: 1, 3, 5, 15

Since 3 is the largest number that appears in both lists, the GCF of 12 and 15 is **3**.

— *or* —

We use the prime factorizations of 12 and 15.

$12 = 2^2 \cdot 3$
$15 = 3 \cdot 5$

The GCF of 12 and 15 is the product of all the prime factors they share. The only prime factor they have in common is 3, so the GCF is **3**.

36. We list the factors of 40 and 56.

40: 1, 2, 4, 5, 8, 10, 20, 40
56: 1, 2, 4, 7, 8, 14, 28, 56

Since 8 is the largest number that appears in both lists, the GCF of 40 and 56 is **8**.

— *or* —

We write the prime factorizations of 40 and 56.

$40 = 2^3 \cdot 5$
$56 = 2^3 \cdot 7$

The GCF of 40 and 56 is the product of all the prime factors they share. So, the GCF is $2^3 = \textbf{8}$.

37. We list the factors of 98 and 168.

98: 1, 2, 7, 14, 49, 98
168: 1, 2, 3, 4, 6, 7, 8, 12, 14, 21, 24, 28, 42, 56, 84, 168

Since 14 is the largest number that appears in both lists, the GCF of 98 and 168 is **14**.

— *or* —

We write the prime factorizations of 98 and 168, grouping any prime factors they have in common.

$98 = 2 \cdot 7^2 = (2 \cdot 7) \cdot 7$
$168 = 2^3 \cdot 3 \cdot 7 = (2 \cdot 7) \cdot 2^2 \cdot 3$

The GCF of 98 and 168 is the product of all the prime factors they share. So, the GCF is $2 \cdot 7 = \textbf{14}$.

38. We list the factors of 63 and 100.

63: 1, 3, 7, 9, 21, 63
100: 1, 2, 4, 5, 10, 20, 25, 50, 100

Since 1 is the largest number that appears in both lists, the GCF of 63 and 100 is **1**.

— *or* —

We write the prime factorizations of 63 and 100, grouping any prime factors they have in common.

$63 = 3^2 \cdot 7$
$100 = 2^2 \cdot 5^2$

63 and 100 do not share any prime factors. However, 1 is a factor of every positive integer, so the GCF is **1**.

39. We write the prime factorizations of 245 and 315, grouping any prime factors they have in common.

$245 = 5 \cdot 7^2 = (5 \cdot 7) \cdot 7$
$315 = 3^2 \cdot 5 \cdot 7 = (5 \cdot 7) \cdot 3^2$

So, the GCF of 245 and 315 is $5 \cdot 7 = \textbf{35}$.

40. We write the prime factorizations of 99 and 495, grouping any prime factors they have in common.

$99 = 3^2 \cdot 11$
$495 = 3^2 \cdot 5 \cdot 11 = (3^2 \cdot 11) \cdot 5$

We see that 99 is a factor of 495. Therefore, the GCF of 99 and 495 is **99**.

41. We write the prime factorizations of 150, 210, and 525.

$150 = 2 \cdot 3 \cdot 5^2$
$210 = 2 \cdot 3 \cdot 5 \cdot 7$
$525 = 3 \cdot 5^2 \cdot 7$

Then, we group prime factors shared by all three numbers:

$150 = 2 \cdot 3 \cdot 5^2 = (3 \cdot 5) \cdot 2 \cdot 5$
$210 = 2 \cdot 3 \cdot 5 \cdot 7 = (3 \cdot 5) \cdot 2 \cdot 7$
$525 = 3 \cdot 5^2 \cdot 7 = (3 \cdot 5) \cdot 5 \cdot 7$

So, $3 \cdot 5 = \textbf{15}$ is the largest integer that is a factor of 150, 210, and 525.

42. Both n and $3n$ have n as a factor. Since n cannot have a factor that is larger than itself, there can be no common factor that is greater than n. So, the GCF of n and $3n$ is $\textbf{\textit{n}}$.

43. We write $12k$ and $18k$ as the product of prime numbers and k.

$12k = 2^2 \cdot 3 \cdot k$
$18k = 2 \cdot 3^2 \cdot k$

Grouping factors that are shared by $12k$ and $18k$ gives:

$12k = (2 \cdot 3 \cdot k) \cdot 2$
$18k = (2 \cdot 3 \cdot k) \cdot 3$

So, for any positive integer k, the GCF of $12k$ and $18k$ is $2 \cdot 3 \cdot k = \textbf{6\textit{k}}$.

44. We write $6r$ and $8r$ as the product of prime numbers and r.

$6r = 2 \cdot 3 \cdot r$
$8r = 2^3 \cdot r$

Grouping factors that are shared by $6r$ and $8r$ gives:

$6r = (2 \cdot r) \cdot 3$
$8r = (2 \cdot r) \cdot 2^2$

So, the GCF of $6r$ and $8r$ is $2r$. Therefore, $2r = 44$. Dividing both sides of this equation by 2 gives $r = \textbf{22}$.

45. The prime factorization of 180 is $2^2 \cdot 3^2 \cdot 5$.
Since $2^2 \cdot 3^2 \cdot 5$ is the GCF of a and b, every common factor of a and b besides 1 is the product of some or all of these primes.

Therefore, to find the second-largest common factor of a and b, we multiply all of their shared prime factors *except one copy of their smallest shared prime, 2.*

So, the second-largest factor a and b have in common is $2 \cdot 3^2 \cdot 5 = $ **90**.

46. Each common factor of 1,008 and 1,620 is either 1 or the product of some or all of the prime factors that these two numbers share. The GCF of 1,008 and 1,620 is the product of *all* of the prime factors that these two numbers share.

Therefore, every common factor of 1,008 and 1,620 is a factor of their GCF. The GCF of 1,008 and 1,620 is 36. So, the common factors of 1,008 and 1,620 are the factors of 36: **1, 2, 3, 4, 6, 9, 12, 18,** and **36.**

*We can use similar reasoning to show that the common factors of **any** two integers are the factors of their GCF!*

FACTORS & MULTIPLES
Euclidean Algorithm   42-45

47. The GCF of any two numbers is the same as the GCF of either number and their difference.

So, GCF(68, 70) = GCF(68, 70−68) = GCF(68, 2).

Since 2 is a factor of 68, we have GCF(68, 2) = 2. So, GCF(68, 70) = GCF(68, 2) = **2.**

48. The GCF of any two numbers is the same as the GCF of either number and their difference.

So, GCF(360, 372) = GCF(360, 372−360) = GCF(360, 12).

Since 12 is a factor of 360, we have GCF(360, 12) = 12. So, GCF(360, 372) = GCF(360, 12) = **12.**

49. We have GCF(70, 91) = GCF(70, 91−70) = GCF(70, 21).

$70 = 2 \cdot 5 \cdot 7$, and $21 = 3 \cdot 7$, so GCF(70, 21) = 7. Therefore, GCF(70, 91) = GCF(70, 21) = **7.**

50. We have GCF(81, 126) = GCF(81, 126−81) = GCF(81, 45).

$81 = 3^4$, and $45 = 3^2 \cdot 5$, so GCF(81, 45) = 3^2 = 9. Therefore, GCF(81, 126) = GCF(81, 45) = **9.**

51. We have GCF(114, 38) = GCF(114−38, 38) = GCF(76, 38).

Since $76 = 38 \cdot 2$, we have GCF(76, 38) = 38. Therefore, GCF(114, 38) = GCF(76, 38) = **38.**

52. GCF(191, 210) = GCF(191, 210−191) = GCF(191, 19).

19 is prime, so its only factors are 1 and 19. Also, since $19 \cdot 10 = 190$, we know 19 is a factor of 190, and 19 is not a factor of 191. Therefore, GCF(191, 19) = 1.

So, GCF(191, 210) = GCF(191, 19) = **1.**

53. Since GCF(a, b) = GCF(a, $a+b$), we have GCF(57, 5643) = GCF(57, 5643+57) = GCF(57, 5700).

Since $5700 = 57 \cdot 100$, we have GCF(57, 5643) = GCF(57, 5700) = **57.**

54. GCF(222, 778) = GCF(222, 222+778) = GCF(222, 1000).

Since 1000 is a power of 10, its only prime factors are 2's and 5's.

$222 = 2 \cdot 111$, and 111 is not divisible by 2 or 5. So, the only prime factor 222 and 1000 share is 2.

Therefore, GCF(222, 778) = GCF(222, 1000) = **2**.

55. If $a = 5$ and $b = 3$, we have GCF(5, 3) = 1. However, GCF(5+3, 5−3) = GCF(8, 2) = 2. So, it is **not always true** that GCF(a, b) is equal to GCF($a+b$, $a−b$).

In fact, we can use any two odd numbers for a and b to show that these GCFs are not the same. Since odd numbers are not divisible by 2, the GCF of any pair of odd numbers cannot have a factor of 2. However, the sum of two odd numbers is even, and the difference between two odd numbers is even, so the GCF of the sum and difference *will* have 2 as a factor.

*Can you find **even** values of a and b that give different results for GCF(a, b) and GCF($a+b$, $a−b$)?*

56. GCF(48, 120) = GCF(48, 120−48) = GCF(48, 72).

We repeat the process of replacing the larger number with the difference between it and the smaller number until we arrive at a GCF that is easy to compute.

GCF(48, 72) = GCF(48, 72−48) = GCF(48, 24). GCF(48, 24) = GCF(48−24, 24) = GCF(24, 24).

So, GCF(48, 120) = GCF(24, 24) = **24.**

57. We repeatedly replace the larger number in the GCF with the difference between it and the smaller number.

GCF(56, 98) = GCF(56, 98−56) = GCF(56, 42). GCF(56, 42) = GCF(56−42, 42) = GCF(14, 42). GCF(14, 42) = GCF(14, 42−14) = GCF(14, 28). GCF(14, 28) = GCF(14, 28−14) = GCF(14, 14).

So, GCF(56, 98) = GCF(14, 14) = **14.**

58. We repeatedly replace the larger number in the GCF with the difference between it and the smaller number.

GCF(216, 126) = GCF(216−126, 126) = GCF(90, 126). GCF(90, 126) = GCF(90, 126−90) = GCF(90, 36). GCF(90, 36) = GCF(90−36, 36) = GCF(54, 36). GCF(54, 36) = GCF(54−36, 36) = GCF(18, 36). GCF(18, 36) = GCF(18, 36−18) = GCF(18, 18).

So, GCF(216, 126) = GCF(18, 18) = **18.**

59. We repeatedly replace the larger number in the GCF with the difference between it and the smaller number.

GCF(245, 385) = GCF(245, 385−245) = GCF(245, 140). GCF(245, 140) = GCF(245−140, 140) = GCF(105, 140). GCF(105, 140) = GCF(105, 140−105) = GCF(105, 35). GCF(105, 35) = GCF(105−35, 35) = GCF(70, 35). GCF(70, 35) = GCF(70−35, 35) = GCF(35, 35).

So, GCF(245, 385) = GCF(35, 35) = **35.**

60. a. We repeatedly replace the larger number with the difference between it and the smaller number until we arrive at a GCF that is easy to compute.

$$\begin{aligned} \text{GCF}(30, 186) &= \text{GCF}(30, 156) \\ &= \text{GCF}(30, 126) \\ &= \text{GCF}(30, 96) \\ &= \text{GCF}(30, 66) \\ &= \text{GCF}(30, 36) \\ &= \text{GCF}(30, 6) \\ &= \textbf{6.} \end{aligned}$$

b. We repeatedly replace the larger number with the difference between it and the smaller number until we arrive at a GCF that is easy to compute.

$$\begin{aligned}
\text{GCF}(360, 51) &= \text{GCF}(309, 51) \\
&= \text{GCF}(258, 51) \\
&= \text{GCF}(207, 51) \\
&= \text{GCF}(156, 51) \\
&= \text{GCF}(105, 51) \\
&= \text{GCF}(54, 51) \\
&= \text{GCF}(3, 51) \\
&= \mathbf{3}.
\end{aligned}$$

c. We repeatedly replace the larger number with the difference between it and the smaller number until we arrive at a GCF that is easy to compute.

$$\begin{aligned}
\text{GCF}(42, 429) &= \text{GCF}(42, 387) \\
&= \text{GCF}(42, 345) \\
&= \text{GCF}(42, 303) \\
&= \text{GCF}(42, 261) \\
&= \text{GCF}(42, 219) \\
&= \text{GCF}(42, 177) \\
&= \text{GCF}(42, 135) \\
&= \text{GCF}(42, 93) \\
&= \text{GCF}(42, 51) \\
&= \text{GCF}(42, 9) \\
&= \mathbf{3}.
\end{aligned}$$

61. Alex repeatedly subtracts 27 from 273 until he gets a number that is 27 or smaller. After subtracting 10 copies of 27, he's left with $273 - 270 = 3$. This is the same thing as computing the remainder of $273 \div 27$!

So, **Alex can use the fact that $273 \div 27$ has remainder 3 to go straight from GCF(273, 27) to GCF(3, 27).**

This same strategy will work any time we use the Euclidean algorithm.

In Problem 60 part (a), repeatedly subtracting 30 from 186 eventually leads to the remainder of $186 \div 30$, which is 6. So, we can go from GCF(30, 186) to GCF(30, 6).

In Problem 60 part (b), since $360 \div 51$ has remainder 3, we can go from GCF(360, 51) to GCF(3, 51).

In Problem 60 part (c), since $429 \div 42$ has remainder 9, we can go from GCF(42, 429) to GCF(42, 9).

In general, we can use the following rule as a shortcut:
If r is the remainder of $a \div b$, then GCF(a, b) = GCF(r, b).

62. Since $42 \cdot 10 = 420$, we have $434 = 42 \cdot 10 + 14$.

So, $434 \div 42$ has remainder 14. In other words, if we repeatedly subtract 42 from 432, we eventually get 14.

Therefore, $\text{GCF}(42, 434) = \text{GCF}(42, 14) = \mathbf{14}$.

63. Since $120 \cdot 20 = 2{,}400$, we have $2{,}424 = 120 \cdot 20 + 24$.

So, repeatedly subtracting 120 from 2,424 eventually gives us 24.

Therefore, $\text{GCF}(120, 2424) = \text{GCF}(120, 24)$. Since $120 = 24 \cdot 5$, we have $\text{GCF}(120, 24) = \mathbf{24}$.

64. Since $90 \cdot 10 = 900$, we have $912 = 90 \cdot 10 + 12$.
So, $912 \div 90$ has remainder 12.

Therefore, $\text{GCF}(90, 912) = \text{GCF}(90, 12)$.

Then, $12 \cdot 7 = 84$, so $90 \div 12$ has remainder 6.
Therefore, $\text{GCF}(90, 12) = \text{GCF}(6, 12)$.

So, $\text{GCF}(90, 912) = \text{GCF}(90, 12) = \text{GCF}(6, 12) = \mathbf{6}$.

65. Since $211 \cdot 5 = 1{,}055$, we have $1{,}085 = 211 \cdot 5 + 30$.
So, $1{,}085 \div 211$ has remainder 30.

Therefore, $\text{GCF}(211, 1085) = \text{GCF}(211, 30)$.

Then, $30 \cdot 7 = 210$, so $211 \div 30$ has remainder 1.
Therefore, $\text{GCF}(211, 30) = \text{GCF}(1, 30)$.

So, $\text{GCF}(211, 1085) = \text{GCF}(211, 30) = \text{GCF}(1, 30) = \mathbf{1}$.

FACTORS & MULTIPLES
Least Common Multiple 46-47

66. We list the first few positive multiples of 12 and 14.

12: 12, 24, 36, 48, 60, 72, (84), 96, ...
14: 14, 28, 42, 56, 70, (84), 98, 112, ...

The smallest number that is in both lists is 84. So, the LCM of 12 and 14 is **84**.

— *or* —

We use the prime factorizations of 12 and 14.

$12 = 2^2 \cdot 3$, so every multiple of 12 has at least two 2's and one 3 in its prime factorization.
$14 = 2 \cdot 7$, so every multiple of 14 has at least one 2 and one 7 in its prime factorization.

So, a multiple of both 12 and 14 must have at least two 2's, one 3, and one 7 in its prime factorization. The least common multiple is the number with only these prime factors: $2^2 \cdot 3 \cdot 7 = \mathbf{84}$.

We have
$84 = 2^2 \cdot 3 \cdot 7 = (2^2 \cdot 3) \cdot 7 = (12) \cdot 7$, and
$84 = 2^2 \cdot 3 \cdot 7 = (2 \cdot 7) \cdot 2 \cdot 3 = (14) \cdot 6$.

67. We list the first few positive multiples of 32 and 40.

32: 32, 64, 96, 128, (160), 192, ...
40: 40, 80, 120, (160), 200, 240, ...

The smallest number that is in both lists is 160. So, the LCM of 32 and 40 is **160**.

— *or* —

We use the prime factorizations of 32 and 40.

$32 = 2^5$, so every multiple of 32 has at least five 2's in its prime factorization.

$40 = 2^3 \cdot 5$, so every multiple of 40 has at least three 2's and one 5 in its prime factorization.

So, a multiple of both 32 and 40 must have at least five 2's and one 5 in its prime factorization. The least common multiple is the number with only these prime factors: $2^5 \cdot 5 = \mathbf{160}$.

68. We use the prime factorizations of 20 and 21.

$20 = 2^2 \cdot 5$, so every multiple of 20 has at least two 2's and one 5 in its prime factorization.

$21 = 3 \cdot 7$, so every multiple of 21 has at least one 3 and one 7 in its prime factorization.

So, a multiple of both 20 and 21 must have at least two 2's, one 3, one 5, and one 7 in its prime factorization. The least common multiple is the number with only these prime factors: $2^2 \cdot 3 \cdot 5 \cdot 7 = $ **420**.

Notice that if we had tried to list multiples to find the LCM, we would have had to write $420 \div 20 = 21$ multiples of 20, and $420 \div 21 = 20$ multiples of 21!

69. We write the prime factorizations of 140 and 160.

$140 = 2^2 \cdot 5 \cdot 7$
$160 = 2^5 \cdot 5$

To compute the LCM, we take the largest power of each prime in either number's prime factorization. So, the LCM of 140 and 160 is $2^5 \cdot 5 \cdot 7 = $ **1,120**.

70. We write the prime factorizations of 28 and 336.

$28 = 2^2 \cdot 7$
$336 = 2^4 \cdot 3 \cdot 7$

We notice that 336 is a multiple of 28:
$336 = 2^4 \cdot 3 \cdot 7 = (2^2 \cdot 7) \cdot 2^2 \cdot 3 = (28) \cdot 2^2 \cdot 3$.

So, the LCM of 28 and 336 is **336**.

71. We write the prime factorizations of 33 and 70.

$33 = 3 \cdot 11$
$70 = 2 \cdot 5 \cdot 7$

To compute the LCM, we take the largest power of each prime in either number's prime factorization. So, the LCM of 33 and 70 is $2 \cdot 3 \cdot 5 \cdot 7 \cdot 11 = $ **2,310**.

72. We write the prime factorization of each integer.

$18 = 2 \cdot 3^2$
$24 = 2^3 \cdot 3$
$30 = 2 \cdot 3 \cdot 5$

We seek the LCM of all three numbers. To compute this, we take the largest power of each prime in any number's prime factorization.

The greatest power of 2 in any prime factorization is 2^3.
The greatest power of 3 in any prime factorization is 3^2.
The greatest power of 5 in any prime factorization is 5^1.

There are no other primes to consider. So, the LCM of 18, 24, and 30 is $2^3 \cdot 3^2 \cdot 5 = $ **360**.

73. We seek the LCM of the first 10 positive integers. Every integer is a multiple of 1. We write the prime factorization of the integers from 2 to 10:

2	5	$8 = 2^3$
3	$6 = 2 \cdot 3$	$9 = 3^2$
$4 = 2^2$	7	$10 = 2 \cdot 5$

The greatest power of 2 in any prime factorization is 2^3.
The greatest power of 3 in any prime factorization is 3^2.
The greatest power of 5 in any prime factorization is 5^1.
The greatest power of 7 in any prime factorization is 7^1.

There are no other primes to consider. So, the LCM of the first 10 positive integers is $2^3 \cdot 3^2 \cdot 5 \cdot 7 = $ **2,520**.

74. Since $28 = 2^2 \cdot 7$ and $35 = 5 \cdot 7$, a multiple of both numbers must have at least two 2's, one 5, and one 7 in its prime factorization. So, every common multiple of 28 and 35 is a multiple of $2^2 \cdot 5 \cdot 7$.

The smallest multiple is the number with only these prime factors: $2^2 \cdot 5 \cdot 7 = 140$.

The second-smallest multiple is then $(2^2 \cdot 5 \cdot 7) \cdot 2 = $ **280**.

75. A number that is divisible by $4 = 2^2$, 5, and $6 = 2 \cdot 3$ must have at least two 2's, one 3, and one 5 in its prime factorization. Such a number can be written as $(2^2 \cdot 3 \cdot 5) \cdot n$, where n is some positive integer.

If $n = 1$, we get $(2^2 \cdot 3 \cdot 5) \cdot 1 = 60$, the LCM of 4, 5, and 6.
If $n = 2$, we get $(2^2 \cdot 3 \cdot 5) \cdot 2 = 120$.
If $n = 3$, we get $(2^2 \cdot 3 \cdot 5) \cdot 3 = 180$.

We see that the numbers that are divisible by 4, 5, and 6 are the multiples of their LCM, 60. So, we count the positive multiples of 60 that are less than 500:

60, 120, 180, 240, 300, 360, 420, 480.

We count 8 multiples, so there are **8** positive integers less than 500 that are divisible by 4, 5, and 6.

76. The multiples of n are n, $2n$, $3n$, $4n$, $5n$, ...
We see that $4n$ is in this list. So, the LCM of n and $4n$ is **$4n$**.

77. The multiples of $6k$ are $6k$, $12k$, $18k$, $24k$, ...
The multiples of $9k$ are $9k$, $18k$, $27k$, $36k$, ...

The smallest multiple that appears in both lists is $18k$. So, the LCM of $6k$ and $9k$ is **$18k$**.

— *or* —

We can write $6k$ as $2 \cdot 3 \cdot k$ and $9k$ as $3^2 \cdot k$.

Since the prime factorizations of $6k$ and $9k$ both include k, the prime factorization of the LCM of $6k$ and $9k$ must include k. So, the LCM of $6k$ and $9k$ is $2 \cdot 3^2 \cdot k = $ **$18k$**.

We have:
$18k = (6k) \cdot 3$, and
$18k = (9k) \cdot 2$.

78. We write $12p$ and $18p$ as the product of prime numbers and p.

$12p = 2^2 \cdot 3 \cdot p$
$18p = 2 \cdot 3^2 \cdot p$

So, the LCM of $12p$ and $18p$ is $2^2 \cdot 3^2 \cdot p = 36p$.
Therefore, $36p = 720$. Dividing both sides of this equation by 36, we have $p = $ **20**.

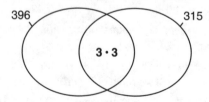

79. We write the prime factorization of each number.

$396 = 2^2 \cdot 3^2 \cdot 11$
$315 = 3^2 \cdot 5 \cdot 7$

We fill in the Venn diagram so that shared prime factors go in the overlapping section. 396 and 315 each have two 3's. So, we fill in the overlapping section as shown:

396   315

$3 \cdot 3$

The remaining prime factors are placed so that the

product of the numbers within the left oval is 396 and the product of the numbers within the right oval is 315.

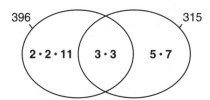

80. The GCF of 396 and 315 is $3^2 = $ **9**.

81. The GCF of 396 and 315 is the product of all of the prime factors they share. **So, the GCF is represented by the product of the primes that are in the overlapping section of the Venn diagram.**

82. We begin with the prime factorizations of 132 and 231.

$132 = 2^2 \cdot 3 \cdot 11$
$231 = 3 \cdot 7 \cdot 11$

We fill in the Venn diagram so that shared prime factors go in the overlapping section. Then, we place the remaining prime factors so that the product within the left oval is 132, and the product within the right oval is 231.

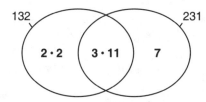

83. The LCM of 132 and 231 is $2^2 \cdot 3 \cdot 7 \cdot 11 = $ **924**.

84. In the previous problem, we learned that the LCM of 132 and 231 is $2^2 \cdot 3 \cdot 7 \cdot 11$. We notice that this is the product of every prime in the two numbers' Venn diagram!

In general, the product of every prime in the Venn diagram of two integers includes the prime factorizations of each integer. So, the product of every prime in the diagram is a common multiple of both integers.

If we remove any prime from this product, then the prime factorization of at least one integer will no longer be included in the product. So, the product of every prime in the diagram is the *least* common multiple of both integers.

Therefore, to compute the LCM of 132 and 231, we compute the product of every prime in their Venn diagram: $(2 \cdot 2) \cdot (3 \cdot 11) \cdot (7) = 924$.

85. We begin with the prime factorizations of 72 and 112.

$72 = 2^3 \cdot 3^2$
$112 = 2^4 \cdot 7$

We fill in the Venn diagram so that shared prime factors go in the overlapping section. Then, we place the remaining prime factors so that the product within the left oval is 72, and the product within the right oval is 112.

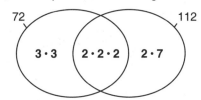

86. The product of the prime factors shared by both numbers is $2 \cdot 2 \cdot 2 = 8$.

The product of every prime in the Venn diagram is $(3 \cdot 3) \cdot (2 \cdot 2 \cdot 2) \cdot (2 \cdot 7) = 1,008$.

So, the GCF of 72 and 112 is **8**, and the LCM of 72 and 112 is **1,008**.

87. The product of 8 and 1,008 is **8,064**.

88. The product of 72 and 112 is **8,064**. This is equal to the product of the GCF and LCM we computed in Problem 87.

The Venn diagram helps explain why the product of two numbers is equal to the product of their GCF and LCM.

When we take the product of two numbers, we multiply every prime in the left oval by every prime in the right oval.

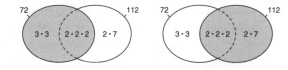

When we take the product of the GCF and LCM, we multiply all of the primes in the overlapping section with all of the primes in the entire diagram.

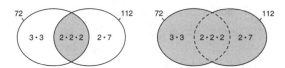

Both products include the same group of primes: every prime in the diagram, along with an extra copy of each prime in the overlapping section.

89. In the previous problem, we learned that the product of two numbers is equal to the product of their GCF and LCM.

In this problem, the product of two numbers is 490. So, the product of their GCF and LCM is also 490. Since their GCF is 7, we have $7 \cdot \text{LCM} = 490$.

Since $7 \cdot 70 = 490$, the LCM of the two numbers is **70**.

One possibility for the two numbers is 7 and 70. Can you think of another pair of numbers with product 490 and GCF 7?

90. The prime factorization of 600 is the product of all of the primes within its oval in the Venn diagram.

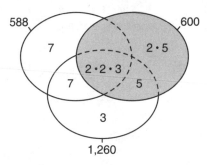

So, the prime factorization of 600 is
$(2 \cdot 2 \cdot 3) \cdot (5) \cdot (2 \cdot 5) = 2^3 \cdot 3 \cdot 5^2$.

91. The GCF of 588, 600, and 1,260 is the product of all the prime factors they share. These are the primes in the section of the Venn diagram where all three ovals overlap.

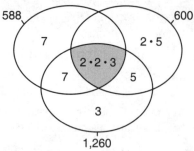

So, the GCF of 588, 600, and 1,260 is $2 \cdot 2 \cdot 3 = \textbf{12}$.

92. If we take the product of every prime in the Venn diagram, the prime factorizations of 588, 600, and 1,260 will be included in that product. So, the product of every prime in the Venn diagram is a common multiple of 588, 600, and 1,260.

If we remove any prime factor from this product, then the prime factorization of at least one of 588, 600, and 1,260 will no longer be included in the product. So, the product of every prime in the diagram is the *least* common multiple of 588, 600, and 1,260.

Therefore, to compute the LCM of 588, 600, and 1,260, we take the product every prime in the diagram:

$(7) \cdot (7) \cdot (2 \cdot 2 \cdot 3) \cdot (3) \cdot (5) \cdot (2 \cdot 5) = \textbf{88,200}$.

93. The GCF of 588 and 1,260 is the product of all of the prime factors that these two numbers share. These are the primes in the section of the Venn diagram where these two numbers' ovals overlap.

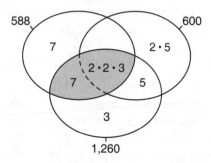

So, the GCF of 588 and 1,260 is $(7) \cdot (2 \cdot 2 \cdot 3) = \textbf{84}$.

94. The LCM of 600 and 1,260 is the product of every prime contained within the two numbers' ovals.

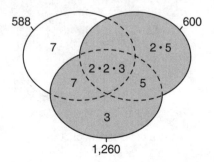

The product of all of the primes in the 1,260 oval is 1,260. We have an extra 2 and 5 in the region inside the 600 oval but outside the 1,260 oval. So, the LCM of 600 and 1,260 is $1,260 \cdot (2 \cdot 5) = 1,260 \cdot 10 = \textbf{12,600}$.

95. We begin with the prime factorization of each number.

$120 = 2^3 \cdot 3 \cdot 5$
$175 = 5^2 \cdot 7$
$210 = 2 \cdot 3 \cdot 5 \cdot 7$

The only prime factor shared by all three numbers is 5. So, we place 5 in the section of the Venn diagram where all three ovals overlap.

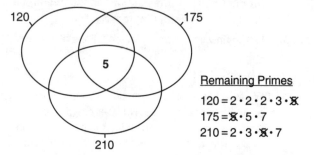

Remaining Primes
$120 = 2 \cdot 2 \cdot 2 \cdot 3 \cdot \cancel{5}$
$175 = \cancel{5} \cdot 5 \cdot 7$
$210 = 2 \cdot 3 \cdot \cancel{5} \cdot 7$

Then, we consider which of the remaining primes are shared by each pair of numbers.

120 and 175 share no remaining primes. So, we leave the section where *only* the 120 and 175 ovals overlap blank.

120 and 210 share one 2 and one 3 among the remaining primes. So, we fill the section where *only* the 120 and 210 ovals overlap with $2 \cdot 3$.

175 and 210 share one 7 among the remaining primes. So, we fill the section where *only* the 175 and 210 ovals overlap with 7.

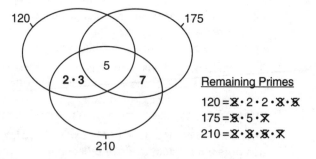

Remaining Primes
$120 = \cancel{2} \cdot 2 \cdot 2 \cdot \cancel{3} \cdot \cancel{5}$
$175 = \cancel{5} \cdot 5 \cdot \cancel{7}$
$210 = \cancel{2} \cdot \cancel{3} \cdot \cancel{5} \cdot \cancel{7}$

Among the primes now remaining, 120 has two 2's it does not share with any other number, and 175 has one 5 it does not share with any other number. So, we complete the diagram by placing the remaining primes as shown below.

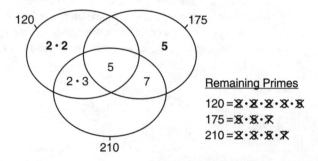

Remaining Primes
$120 = \cancel{2} \cdot \cancel{2} \cdot \cancel{2} \cdot \cancel{3} \cdot \cancel{5}$
$175 = \cancel{5} \cdot \cancel{5} \cdot \cancel{7}$
$210 = \cancel{2} \cdot \cancel{3} \cdot \cancel{5} \cdot \cancel{7}$

96. We use the diagram we made in Problem 95.

The only prime factor shared by all three numbers is 5.

The product of every prime in the Venn diagram is $(2 \cdot 2) \cdot (2 \cdot 3) \cdot (5) \cdot (7) \cdot (5) = 4,200$.

So, the GCF of 120, 175, and 210 is **5** and the LCM of 120, 175, and 210 is **4,200**.

*Note that when we have three numbers, the product of all three numbers is **not** equal to the product of their GCF and LCM!*

FACTORS & MULTIPLES
Word Problems 51-52

97. We list the number of seconds at which each worlump croaks.

Baby: 9, 18, 27, 36, ㊺
Adult: 15, 30, ㊺

They croak at the same time in **45** seconds.

98. Patties can only be bought in multiples of 8, and buns in multiples of 6. We're told that Cruggle buys an equal number of patties and buns, so the number of each item that he buys is a common multiple of 8 and 6.

The least common multiple of 8 and 6 is 24, so the smallest number of buns Cruggle could have bought is 24. Since buns come in packages of 6, the smallest number of packages of buns he could have bought is $24 \div 6 = $ **4**.

99. Ren has \$126 worth of snargles. So, the value of each snargle must be a factor of 126.

Ben has \$182 worth of snargles. So, the value of each snargle must also be a factor of 182.

We have $126 = 2 \cdot 3^2 \cdot 7$ and $182 = 2 \cdot 7 \cdot 13$. So, the greatest common factor of 126 and 182 is $2 \cdot 7 = 14$. Therefore, the largest possible value of a snargle is **\$14**.

100. Gardina distributes all 40 apples evenly into the baskets. So, the number of baskets must be a factor of 40.

Similarly, Gardina has 28 oranges, so the number of baskets is a factor of 28. She has 36 bananas, so the number of baskets is a factor of 36.

The greatest common factor of 40, 28, and 36 is 4. So, the largest number of baskets Gardina can make is 4.

Each of the 4 baskets will have $40 \div 4 = 10$ apples, $28 \div 4 = 7$ oranges, and $36 \div 4 = 9$ bananas.

101. Byath Leet runs every 6 days and swims every 10 days, so the number of days from each super-day to the next is the least common multiple of 6 and 10, which is 30.

Since Byath Leet's 10th super-day is today, he has $20 - 10 = 10$ more super-days to get to his 20th super-day. There are 30 days from one super-day to the next, so it will be $10 \cdot 30 = $ **300** days until he has his 20th super-day.

102. One ninth of the gymnasts at Acrobeast Gymnasium have a tail. We can only take one ninth of a number and get an integer result if that number is divisible by 9. So, the number of gymnasts at Acrobeast Gymnasium is a multiple of 9.

Similarly, since one twelfth of the gymnasts have fur, the number of gymnasts must be a multiple of 12, and since one fifteenth of the gymnasts have webbed feet, the number of gymnasts must be a multiple of 15.

The smallest number that is a multiple of 9, 12, and 15 is the LCM of those numbers. Since $9 = 3^2$, $12 = 2^2 \cdot 3$, and $15 = 3 \cdot 5$, their LCM is $2^2 \cdot 3^2 \cdot 5 = 180$.

So, the smallest number of gymnasts that could attend Acrobeast Gymnasium is **180**.

103. There are 207 stairs between the bottom floor and the pool floor. So, the number of stairs between consecutive floors is a factor of 207.

There are 368 stairs between the bottom floor and the top floor. So, the number of stairs between consecutive floors is also a factor of 368.

We have $207 = 3^2 \cdot 23$ and $368 = 2^4 \cdot 23$. The only prime that 207 and 368 share is 23. Every integer has 1 as a factor. So, the only common factors of 207 and 368 are 23 and 1. Having just 1 stair between floors is unreasonable, so there must be 23 stairs between each floor.

We know there are 207 stairs between the bottom floor (which is floor 1) and the pool floor. So, there are $207 \div 23 = 9$ sets of stairs between the floor 1 and the pool floor. Each set of stairs goes up one floor, so the pool is on floor $1 + 9 = $ **10**.

104. Since Niki ends the game with 0 points, the number of points she gains must equal the number of points she loses. The points Niki gains is a multiple of 45, and the points Niki loses is a multiple of 12. So, we seek a common multiple of 45 and 12.

Since we want the smallest number of rolls she could have made, we seek the *least* common multiple of 45 and 12.

$45 = 3^2 \cdot 5$, and $12 = 2^2 \cdot 3$, so LCM$(45, 12) = 2^2 \cdot 3^2 \cdot 5 = 180$.

So, 180 is the smallest number of points Niki could have gained and lost to end up with 0 points. To gain 180 points, Niki must roll the die $180 \div 45 = 4$ times. To lose 180 points, Niki must roll the die $180 \div 12 = 15$ times.

Therefore, the smallest number of rolls Niki could make to end with 0 points is $4 + 15 = $ **19**.

FACTORS & MULTIPLES
GCF-LCM Webs 53-55

105. We write the prime factorizations of 20 and 32.

$20 = 2^2 \cdot 5$
$32 = 2^5$

So, GCF$(20, 32) = 2^2 = $ **4**, and LCM$(20, 32) = 2^5 \cdot 5 = $ **160**. We fill the circles as shown below.

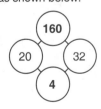

106. We write the prime factorization of each number, and label the missing number x, as shown below.

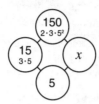

The LCM of 15 and x is $150 = 2 \cdot 3 \cdot 5^2$. The 2 and the 5^2 in this LCM do not come from $15 = 3 \cdot 5$, so they must come from x. So, x includes exactly one 2 and exactly two 5's in its prime factorization.

If the prime factorization of x included any primes other than 2, 3, or 5, then the LCM of 15 and x would also include these primes. So, the prime factorization of x does not include any primes other than 2, 3, or 5.

Finally, if the prime factorization of x included any 3's, then the GCF of 15 and x would have 3 as a factor. So, the prime factorization of x does not include any 3's.

Therefore, the value of x is $2 \cdot 5^2 = \mathbf{50}$.

— *or* —

In the Venn Diagram section of this chapter, we learned that the product of two integers is equal to the product of their GCF and LCM. The GCF of 15 and x is 5, and the LCM of 15 and x is 150. Therefore, $15x = 5 \cdot 150 = 750$.

Dividing both sides of $15x = 750$ by 15 gives $x = \mathbf{50}$.

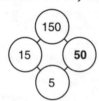

107. We write the prime factorization of each number and label the missing number x, as shown below.

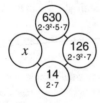

The LCM of 126 and x is $630 = 2 \cdot 3^2 \cdot 5 \cdot 7$. The 5 in this LCM does not come from $126 = 2 \cdot 3^2 \cdot 7$, so it must come from x. So, the prime factorization of x has exactly one 5.

The GCF of 126 and x is $14 = 2 \cdot 7$. So, the prime factorization of x includes at least one 2 and at least one 7. If the prime factorization of x included more than one 2 or one 7, then the LCM of x and 126 would also include more than one 2 or 7. So, the prime factorization of x includes *exactly* one 2 and one 7.

If the prime factorization of x included any prime factors other than $2 \cdot 5 \cdot 7$, then the LCM of x and 126 would be greater than 630. Therefore, $x = 2 \cdot 5 \cdot 7 = \mathbf{70}$.

— *or* —

The product of two integers is equal to the product of their GCF and LCM. So, $126x = 14 \cdot 630$. Instead of computing $14 \cdot 630$ and then dividing both sides of the equation by 126, we rewrite the equation using prime factors:

$$(2 \cdot 3^2 \cdot 7) \cdot x = (2 \cdot 7) \cdot (2 \cdot 3^2 \cdot 5 \cdot 7).$$

Rearranging primes on the equation's right side gives

$$(2 \cdot 3^2 \cdot 7) \cdot x = (2 \cdot 7) \cdot (2 \cdot 3^2 \cdot 5 \cdot 7)$$
$$= (2 \cdot 3^2 \cdot 7) \cdot (2 \cdot 5 \cdot 7).$$

This equation is true when $x = 2 \cdot 5 \cdot 7 = \mathbf{70}$.

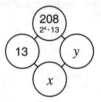

108. We write the prime factorization of 208, and label the two blank circles x and y, as shown below.

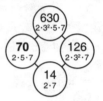

The GCF of 13 and y is x. Since 13 is prime, x can only be 1 or 13. However, no two circles in a web can have the same value. So, x cannot be 13 and is therefore **1**.

Then, $208 = 2^4 \cdot 13$ is the LCM of 13 and y. So, y must be $2^4 = 16$ or $2^4 \cdot 13 = 208$. However, no two circles can have the same value. So, $y = 2^4 = \mathbf{16}$.

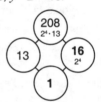

109. GCF(10, 4) = **2**, and LCM(10, 4) = **20**. We fill the circles as shown below.

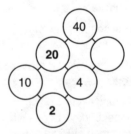

Then, we write the prime factorizations of 4, 20, and 40, and label the number in the remaining blank x.

The LCM of 20 and x is $40 = 2^3 \cdot 5$. The 2^3 in this LCM does not come from $20 = 2^2 \cdot 5$, so it must come from x. So, the prime factorization of x has exactly three 2's.

If the prime factorization of x included any primes other than 2 or 5, then the LCM of 20 and x would also include those primes. Also, if the prime factorization of x included a 5, then the GCF of 20 and x would have a factor of 5.

So, 2 is the only prime in the prime factorization of x. Therefore, $x = 2^3 = \mathbf{8}$.

— *or* —

We fill the first two circles as in the previous solution.

Then, the product of 20 and x is equal to the product of their GCF and LCM. So, $20x = 4 \cdot 40 = 160$.

Dividing both sides of $20x = 160$ by 20 gives $x = \mathbf{8}$.

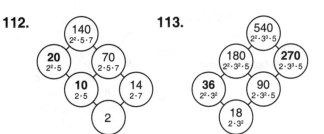

We use the strategies discussed in the previous problems to complete the following webs.

110.
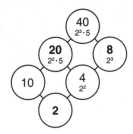

111.
225 web: 990 (2·3²·5·11), 330 (2·3·5·11), 198 (2·3²·11), 110 (2·5·11), 66 (2·3·11), 22 (2·11)

112.
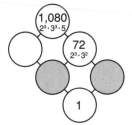

113.
540 (2²·3³·5), 180 (2²·3²·5), 270 (2·3³·5), 36 (2²·3²), 90 (2·3²·5), 18 (2·3²)

114. We write the prime factorization of each given number, then consider the shaded circles below.

1,080 ($2^3 \cdot 3^3 \cdot 5$) / 72 ($2^3 \cdot 3^2$) / 1

The LCM of these shaded circles is $72 = 2^3 \cdot 3^2$, and their GCF is 1. So, they share no prime factors. The only pairs

of numbers with LCM 72 and GCF 1 are (72 and 1) and (8 and 9). However, 72 and 1 already appear in the web. Therefore, these two circles must be filled with 8 and 9.

We label the number in the top blank circle x, as shown.

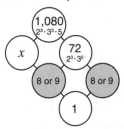

Since the prime factorization of 1,080 has three 3's, and the prime factorization of 72 has only two 3's, we know the prime factorization of x has exactly three 3's. So, the GCF of x and 72 has two 3's in its prime factorization.

Therefore, the left shaded circle is **9**, and the right shaded circle is **8**.

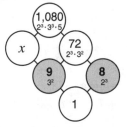

Then, the product of 72 and x is equal to the product of their GCF and LCM. So, $72x = 9 \cdot 1{,}080$. Rewriting this equation using prime factors gives

$$(2^3 \cdot 3^2) \cdot x = (3^2) \cdot (2^3 \cdot 3^3 \cdot 5)$$
$$= (2^3 \cdot 3^2) \cdot (3^3 \cdot 5).$$

This equation is true when $x = 3^3 \cdot 5 = \mathbf{135}$.

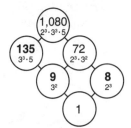

115. We write the prime factorization of each given number and label the left blank circle x, as shown below.

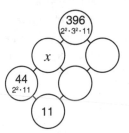

Since x is the LCM of $44 = 2^2 \cdot 11$ and some other integer, the prime factorization of x must have at least two 2's and at least one 11. Since $396 = 2^2 \cdot 3^2 \cdot 11$ is the LCM of x and some other integer, the prime factorization of x cannot have more than two 2's or more than one 11.

So, the prime factorization of x includes exactly two 2's and one 11.

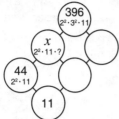

However, x cannot equal 44, so 2 and 11 cannot be its only prime factors. Since $396 = 2^2 \cdot 3^2 \cdot 11$ is the LCM of x and another integer, the only other prime factor x could have is 3. Since x cannot equal 396, it must have exactly one 3 in its prime factorization.

So, $x = 2^2 \cdot 3 \cdot 11 = \textbf{132}$.

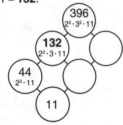

We use the strategies discussed in previous problems to fill the remaining blank circles.

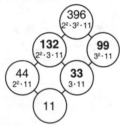

116. We write the prime factorization of each given number, and label the bottom blank circle x, as shown below.

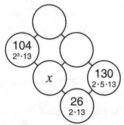

The GCF of x and 130 is $26 = 2 \cdot 13$. So, the prime factorization of x has at least one 2 and at least one 13.

Then, since x is the GCF of 104 and some other integer, and 104 has only 2 and 13 as prime factors, x cannot have any prime factors other than 2 and 13.

Finally, x cannot equal $26 = 2 \cdot 13$ or $104 = 2^3 \cdot 13$, so its only possible value is $2^2 \cdot 13 = \textbf{52}$.

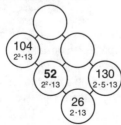

We fill the remaining blank circles as shown below.

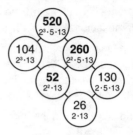

We use the strategies discussed in the previous problems to complete the following webs.

117.

118.

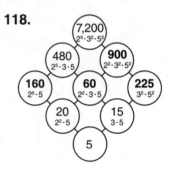

119. We label the bottom blank circle x, as shown to the right.

Since x is the GCF of 25 and some other integer, x must be a factor of 25 (1, 5, or 25). Both 1 and 25 already appear in the web, so $x = \textbf{5}$.

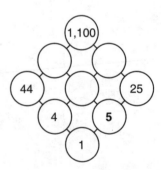

We use the strategies discussed in previous problems to fill the remaining blank circles.

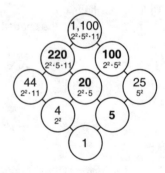

120. Using strategies discussed in previous problems, we fill in the top-right blank circle as shown below.

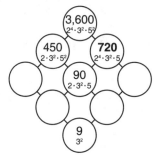

Next, we consider the two bottom blank circles. The numbers in these circles have GCF $9 = 3^2$ and LCM $90 = 2 \cdot 3^2 \cdot 5$. So, each has exactly two 3's in its prime factorization. Since these two numbers must be different, one of them is $2 \cdot 3^2 = 18$, and the other is $3^2 \cdot 5 = 45$.

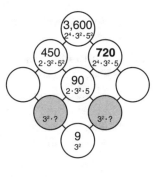

We label the number in the leftmost blank y. The LCM of 90 and y is 450. Since 450 has two 5's in its prime factorization and 90 has only one 5, the prime factorization of y has exactly two 5's. Therefore, the GCF of 90 and y has exactly one 5 in its prime factorization.

So, the bottom-left blank is **45**, and the bottom-right blank is **18**.

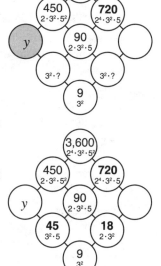

We use strategies discussed in previous problems to solve for the remaining blanks.

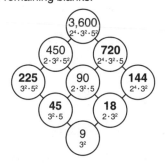

121. The expression 7! means $7 \cdot 6 \cdot 5 \cdot 4 \cdot 3 \cdot 2 \cdot 1$. Writing each composite factor as a product of prime factors, then grouping like primes, we have

$$7! = 7 \cdot 6 \cdot 5 \cdot 4 \cdot 3 \cdot 2 \cdot 1$$
$$= 7 \cdot (2 \cdot 3) \cdot 5 \cdot (2 \cdot 2) \cdot 3 \cdot 2$$
$$= 2^4 \cdot 3^2 \cdot 5 \cdot 7.$$

So, the power of 2 in the prime factorization of 7! is 2^4.

— *or* —

There are three multiples of 2 from 1 to 7, each of which contributes a 2 to the prime factorization of (7!).

$$7! = 7 \cdot \textcircled{6} \cdot 5 \cdot \textcircled{4} \cdot 3 \cdot \textcircled{2} \cdot 1$$

However, $4 = 2^2$ has two 2's in its prime factorization, so it contributes an additional factor of 2. Therefore, there are $3 + 1 = 4$ twos in the prime factorization of (7!). The power of 2 in 7! is 2^4.

122. The expression 18! means $18 \cdot 17 \cdot 16 \cdot \dots \cdot 3 \cdot 2 \cdot 1$.

There are 6 multiples of 3 from 1 to 18, each of which contributes a 3 to the prime factorization of (18!).

Then, every multiple of $9 = 3^2$ contributes a factor of 3 that we have not yet counted. There are 2 multiples of 9 from 1 to 18. So, we have two additional factors of 3.

Since $3^3 = 27$ is greater than 18, we do not have any more factors of 3 to count.

Therefore, there are $6 + 2 = 8$ threes in the prime factorization of (18!). The power of 3 in 18! is 3^8.

123. We consider the number of 2's in the prime factorization of each factorial.

4! is the product of the integers from 1 to 4.

There are 2 multiples of 2 from 1 to 4, each contributing a factor of 2 to the prime factorization of (4!).

There is 1 multiple of $4 = 2^2$ from 1 to 4, which contributes a factor of 2 we have not yet counted.

Therefore, there are $2 + 1 = 3$ twos in the prime factorization of (4!). The power of 2 in 4! is 2^3.

8! is the product of the integers from 1 to 8.

There are 4 multiples of 2 from 1 to 8, each contributing a factor of 2 to the prime factorization of (8!).

There are 2 multiples of $4 = 2^2$ from 1 to 8, each contributing a factor of 2 that we have not yet counted.

There is 1 multiple of $8 = 2^3$ from 1 to 8, which contributes a factor of 2 that we have not yet counted.

Therefore, there are $4 + 2 + 1 = 7$ twos in the prime factorization of (8!). The power of 2 in 8! is 2^7.

16! is the product of the integers from 1 to 16.

There are 8 multiples of 2 from 1 to 16, each contributing a factor of 2 to the prime factorization of (16!).

There are 4 multiples of $4 = 2^2$ from 1 to 16, each contributing a factor of 2 that we have not yet counted.

There are 2 multiples of $8 = 2^3$ from 1 to 16, each contributing a factor of 2 that we have not yet counted.

There is 1 multiple of $16 = 2^4$ from 1 to 16, which contributes a factor of 2 that we have not yet counted.

Therefore, there are $8+4+2+1 = 15$ twos in the prime factorization of (16!). The power of 2 in 16! is **2^{15}**.

<u>32!</u> is the product of the integers from 1 to 32.

There are 16 multiples of 2 from 1 to 32.
There are 8 multiples of $4 = 2^2$ from 1 to 32.
There are 4 multiples of $8 = 2^3$ from 1 to 32.
There are 2 multiples of $16 = 2^4$ from 1 to 32.
There is 1 multiple of $32 = 2^5$ from 1 to 32.

Therefore, there are $16+8+4+2+1 = 31$ twos in the prime factorization of (32!). The power of 2 in 32! is **2^{31}**.

Do you notice a pattern? As a challenge, see if you can write an expression for the power of 2 in the prime factorization of (2^n!).

124. Since $6 = 2 \cdot 3$, we seek the largest power of $(2 \cdot 3)$ that is a factor of (12!). The prime factorization of 12! is

$12! = 12 \cdot 11 \cdot 10 \cdot 9 \cdot 8 \cdot 7 \cdot 6 \cdot 5 \cdot 4 \cdot 3 \cdot 2 \cdot 1$
$= (2^2 \cdot 3) \cdot 11 \cdot (2 \cdot 5) \cdot (3^2) \cdot (2^3) \cdot 7 \cdot (2 \cdot 3) \cdot 5 \cdot (2^2) \cdot 3 \cdot 2$
$= 2^{10} \cdot 3^5 \cdot 5^2 \cdot 7 \cdot 11$.

There are ten 2's and five 3's in this prime factorization. So, we can pair five 2's and five 3's to create a factor of $(2 \cdot 3)^5 = 6^5$:

$12! = 2^{10} \cdot 3^5 \cdot 5^2 \cdot 7 \cdot 11$
$= (2 \cdot 3) \cdot (2 \cdot 3) \cdot (2 \cdot 3) \cdot (2 \cdot 3) \cdot (2 \cdot 3) \cdot 2^5 \cdot 5^2 \cdot 7 \cdot 11$
$= (2 \cdot 3)^5 \cdot 2^5 \cdot 5^2 \cdot 7 \cdot 11$
$= (6^5) \cdot 2^5 \cdot 5^2 \cdot 7 \cdot 11$.

There are no more 3's that we can pair with a 2 to create a larger power of 6. So, the largest power of 6 that is a factor of 12! is **6^5**.

— *or* —

Since $6 = 2 \cdot 3$, we seek the largest power of $(2 \cdot 3)$ that is a factor of (12!).

There are fewer multiples of 3 from 1 to 12 than there are multiples of 2. Therefore, there are fewer 3's than 2's in the prime factorization of (12!).

So, the largest power of $(2 \cdot 3)$ that is a factor of 12! is limited by the number of 3's in the prime factorization of (12!). Therefore, we only need to count the number of 3's.

There are 4 multiples of 3 from 1 to 12.
There is 1 multiple of $9 = 3^2$ from 1 to 12.

All together, there are $4+1 = 5$ threes in the prime factorization of (12!).

So, we can pair five 3's with five 2's in the prime factorization of (12!). Therefore, $(2 \cdot 3)^5 = $ **6^5** is the largest power of 6 that is a factor of (12!).

125. 25! is the product of the integers from 1 to 25.

There are 12 multiples of 2 from 1 to 25.
There are 6 multiples of $4 = 2^2$ from 1 to 25.
There are 3 multiples of $8 = 2^3$ from 1 to 25.
There is 1 multiple of $16 = 2^4$ from 1 to 25.
$2^5 = 32$ is greater than 25, so there are no multiples of 2^5 from 1 to 25.

So, the exponent of 2 in the prime factorization of 25! is $12+6+3+1 = 22$. Therefore, the largest power of 2 that is a factor of 25! is **2^{22}**.

126. There are 5 multiples of 5 from 1 to 25.
There is 1 multiple of $25 = 5^2$ from 1 to 25.

So, the exponent of 5 in the prime factorization of 25! is $5+1 = 6$. Therefore, the largest power of 5 that is a factor of 25! is **5^6**.

127. Since $10 = 2 \cdot 5$, we seek the largest power of $(2 \cdot 5)$ that is a factor of (25!).

We showed in the previous two problems that there are twenty-two 2's and six 5's in the prime factorization of (25!). So, we can pair six 2's with six 5's to create a factor of $(2 \cdot 5)^6 = 10^6$.

There are no more 5's we can pair with a 2 to make a larger power of 10. So, the largest power of 10 that is a factor of 25! is **10^6**.

128. Every integer with trailing zeros can be written as the product of an integer without trailing zeros and a power of 10. For example, $700 = 7 \cdot 100 = 7 \cdot 10^2$, and $3,450,000 = 345 \cdot 10,000 = 345 \cdot 10^4$.

We see that the number of trailing zeros an integer has is determined by the largest power of 10 that is a factor of that integer.

In the previous problem, we showed that the largest power of 10 that is a factor of 25! is 10^6. So, 25! has **6** trailing zeros.

129. To find the number of trailing zeros in 100!, we look for the largest power of 10 that is a factor of (100!). This is the same as finding the largest power of $(2 \cdot 5)$ that is a factor of (100!).

There are fewer 5's than 2's in the prime factorization of (100!). So, the largest power of $(2 \cdot 5)$ that is a factor of 100! is limited by the number of 5's in the prime factorization of (100!). Therefore, we only need to count the number of 5's.

There are 20 multiples of 5 from 1 to 100.
There are 4 multiples of $25 = 5^2$ from 1 to 100.
$5^3 = 125$ is greater than 100, so there are no multiples of 5^3 from 1 to 100.

So, the prime factorization of 100! has $20+4 = 24$ fives. We can pair these 24 fives with 24 twos. Therefore, $2^{24} \cdot 5^{24} = (2 \cdot 5)^{24} = 10^{24}$ is the largest power of 10 that is a factor of (100!), giving **24** trailing zeros.

FACTORS & MULTIPLES

Relatively Prime   58-60

130. 7 is prime. So, any number without a 7 in its prime factorization is relatively prime to 7.

We write the prime factorization of each number, and circle those that are relatively prime to 7.

7 (9 = 3²) (15 = 3·5) 28 = 2²·7 (75 = 3·5²)

131. The prime factorization of 45 is $3^2 \cdot 5$. So, any number with no 3's or 5's in its prime factorization is relatively prime to 45.

We write the prime factorization of each number, and circle those that are relatively prime to 45.

(8 = 2³) 18 = 2·3² 21 = 3·7 (44 = 2²·11) 80 = 2⁴·5

132. We write the prime factorization of each number.

$$35 = 5 \cdot 7 \qquad 27 = 3^3 \qquad 15 = 3 \cdot 5$$
$$6 = 2 \cdot 3 \qquad 8 = 2^3 \qquad 22 = 2 \cdot 11$$

35 shares a 5 with 15, but shares no prime factors with 27, 6, 8, or 22. So, we connect 35 with 27, 6, 8, and 22.

Continuing counter-clockwise around the diagram, we compare each number with the numbers in the diagram we have not yet considered.

27 shares a 3 with 15 and 6, but shares no prime factors with 8 or 22. So, we connect 27 with 8 and 22.

15 shares a 3 with 6, but shares no prime factors with 8 or 22. So, we connect 15 with 8 and 22.

6 shares a 2 with 8 and 22, so we draw no additional lines from 6.

8 shares a 2 with 22, so we draw no additional lines from 8.

We have already considered all possible pairings with 22. The completed diagram is shown below.

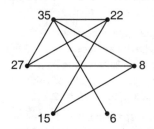

133. The prime factorization of 20 is $2^2 \cdot 5$. So, 20 is relatively prime to any integer without a 2 or a 5 in its prime factorization. This is only true of odd integers that do not have units digit 5. So, the list of positive integers less than 20 that are relatively prime to 20 is:

1, 3, 7, 9, 11, 13, 17, 19.

134. The prime factorization of 210 is $2 \cdot 3 \cdot 5 \cdot 7$. Since the number we seek is greater than 1, it can be written as the product of prime factors, none of which are 2, 3, 5, or 7.

Since 2, 3, 5, and 7 are the first four prime numbers, the prime factorization of the number we seek must include a prime that is greater than 7. The smallest prime that is greater than 7 is 11. Therefore, **11** is the smallest number other than 1 that is relatively prime to 210.

135. We write the prime factorizations of 50 and 70.

$$50 = 2 \cdot 5^2$$
$$70 = 2 \cdot 5 \cdot 7$$

So, any number that has a 2, 5, or 7 in its prime factorization is not relatively prime to both 50 and 70. We can quickly rule out numbers between 50 and 70 that are divisible by 2 or 5. This leaves the following integers:

51, 53, 57, 59, 61, 63, 67, 69.

Of these numbers, only 63 is divisible by 7, so we remove it from the list. The remaining numbers are not divisible by 2, 5, or 7. Therefore, they share no prime factors with 50 or 70 and are relatively prime to both.

51, 53, 57, 59, 61, 67, 69.

There are **7** numbers in this list.

136. Ralph is **incorrect**. For example, let $a = 7$ and $b = 9$. While 7 and 9 are relatively prime, $7 + 1 = 8$ and $9 + 1 = 10$ are not relatively prime, since they share a factor of 2.

You may have come up with a different example to prove that Ralph's statement is incorrect.

137. Consider two relatively prime numbers a and b. The LCM of a and b is a multiple of a, so the prime factorization of the LCM contains all of a's prime factors. Similarly, the LCM is a multiple of b, so the prime factorization of the LCM contains all of b's prime factors.

Since a and b do not share any prime factors, LCM(a, b) is the product of all of a's prime factors and all of b's prime factors. The product of a and b is also the product of all of a's prime factors and all of b's prime factors. So, Cammie's statement is **correct**.

As an example, $28 = 2^2 \cdot 7$ and $45 = 3^2 \cdot 5$ are two relatively prime numbers, and the LCM of 28 and 45 is $2^2 \cdot 3^2 \cdot 5 \cdot 7 = (2^2 \cdot 7) \cdot (3^2 \cdot 5) = 28 \cdot 45$.

— *or* —

Relatively prime numbers share no prime factors. So, the overlapping section of the Venn diagram of their primes is empty.

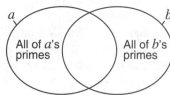

We learned previously that the LCM of two numbers is equal to the product of every prime in their Venn diagram. When two numbers are relatively prime, this is the product of the two numbers. So, Cammie is **correct**.

— *or* —

We learned previously that the product of two numbers is equal to the product of their GCF and LCM. If two numbers are relatively prime, then their GCF is 1. Therefore, their product is equal to 1 times their LCM, which is their LCM.

So, Cammie's statement is **correct**.

138. a. The factors of 9 are 1, 3, and 9.
The factors of 10 are 1, 2, 5, and 10.

So, GCF(9, 10) = **1**.

b. The prime factorization of 65 is 5 · 13. The prime factorization of 66 is 2 · 3 · 11. Since 65 and 66 share no primes in their prime factorization, their only common factor is 1.

So, GCF(65, 66) = **1**.

c. By the Euclidean algorithm, we have
GCF(680, 681) = GCF(680, 681−680) = GCF(680, 1) = **1**.

139. Using the Euclidean algorithm, we have
GCF(n, n+1) = GCF(n, (n+1)−n) = GCF(n, 1).

Since the GCF of 1 and any integer is 1, we have
GCF(n, n+1) = GCF(n, 1) = 1.

Therefore, **if n is a positive integer, then n and n+1 are relatively prime**.

— *or* —

We consider the factors of n. Since n+1 is one more than n, dividing n+1 by any of n's factors leaves a remainder of 1. Therefore, n+1 is not divisible by any of n's factors except for 1.

So, the GCF of n and n+1 is 1. Therefore, **if n is a positive integer, then n and n+1 are relatively prime**.

140. a. The difference between a and b will be as small as possible when a and b are as close together as possible. The closest integers with sum 1,000 are 500 and 500. However, 500 and 500 are not relatively prime.

The next-closest integers with sum 1,000 are 501 and 499. By the Euclidean algorithm, we have:

GCF(501, 499) = GCF(501−499, 499) = GCF(2, 499) = 1.

So, 501 and 499 are relatively prime. Letting a = 501 and b = 499, the smallest possible value of a−b is 501−499 = **2**.

b. The difference between a and b will be as large as possible when a and b are as far apart as possible. Since a and b are both greater than 1, the two integers with sum 1,000 that are farthest apart are 2 and 998. However, 2 and 998 are both divisible by 2, so they are not relatively prime.

The two integers with sum 1,000 that are next-farthest apart are 3 and 997. Since 3 is prime, and 997 is not divisible by 3, these integers are relatively prime.

So, letting a = 997 and b = 3, the largest possible value for a−b is 997−3 = **994**.

141. Since 3 is prime, any integer that is not a multiple of 3 is relatively prime to 3.

We consider three consecutive positive integers on the number line.

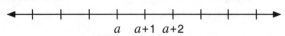

Every third number after a multiple of 3 is also a multiple of 3. So, we have three different possibilities for the numbers that are multiples of 3:

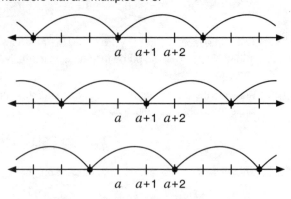

In each case, exactly one of the three consecutive integers is a multiple of three, and the other two are not. So, among any three consecutive positive integers, **two** of them are relatively prime to 3.

142. The prime factorization of 2,880 is $2^6 \cdot 3^2 \cdot 5$. Pairing 2's to make 4's, we have

$$2{,}880 = 2^6 \cdot 3^2 \cdot 5$$
$$= (2 \cdot 2) \cdot (2 \cdot 2) \cdot (2 \cdot 2) \cdot 3^2 \cdot 5$$
$$= (2 \cdot 2)^3 \cdot 3^2 \cdot 5$$
$$= 4^3 \cdot 3^2 \cdot 5.$$

So, the largest power of 4 that is a factor of 2,880 is 4^3.

143. 12,222 is divisible by 6. So, 12,222 plus any multiple of 6 is also divisible by 6. Similarly, 12,222 plus any multiple of 9 is divisible by 9, and 12,222 plus any multiple of 21 is divisible by 21.

So, 12,222 plus a common multiple of 6, 9, and 21 will be divisible by each of 6, 9 and 21. The least common multiple of 6, 9, and 21 is $2 \cdot 3^2 \cdot 7 = 126$. Therefore, the next-largest integer after 12,222 that is divisible by 6, 9, and 21 is 12,222+126 = **12,348**.

144. We use a Venn diagram. Since the GCF of the two integers is 6, we fill the overlapping section of the diagram with 2 · 3.

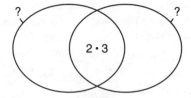

The LCM of the two integers is $72 = 2^3 \cdot 3^2$. We learned previously that the LCM of two integers is the product of every prime in their Venn diagram. So, the remaining sections of the diagram must together contain two 2's and one 3. Also, since the overlapping section is already filled in, the remaining sections cannot contain any of the same primes.

The only ways to place two 2's and one 3 so that the

remaining sections share no primes are shown below. The left and right ovals of each diagram can be switched to give the same pair of numbers.

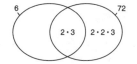 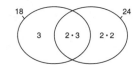

So, the two pairs of integers with a GCF of 6 and an LCM of 72 are **(6 and 72)** and **(18 and 24)**.

— or —

The LCM of the two integers is $72 = \boxed{2^3} \cdot 3^2$, and the GCF of the two integers is $6 = \boxed{2} \cdot 3$. So, one integer has exactly three 2's in its prime factorization, and the other integer has exactly one 2 in its prime factorization.

Similarly, since the LCM is $2^3 \cdot \boxed{3^2}$ and the GCF is $2 \cdot \boxed{3}$, one integer has exactly two 3's in its prime factorization and the other has exactly one 3 in its prime factorization.

So, the integer with three 2's in its prime factorization must have either one or two 3's in its prime factorization.

- If it has one 3, then it is $2^3 \cdot 3 = 24$, and the other integer is $2 \cdot 3^2 = 18$.

- If it has two 3's, then it is $2^3 \cdot 3^2 = 72$, and the other integer is $2 \cdot 3 = 6$.

So, the two pairs of integers with a GCF of 6 and an LCM of 72 are **(6 and 72)** and **(18 and 24)**.

— or —

We learned previously that the product of two integers is equal to the product of their GCF and LCM. The two integers in this problem have GCF 6 and LCM 72. So, their product is $6 \cdot 72 = 432$.

Since the GCF of the two integers is 6, each integer is a multiple of 6. The only factor pairs of 432 for which each factor is a multiple of 6 are $(6 \cdot 72)$, $(12 \cdot 36)$, and $(18 \cdot 24)$.

Of these factor pairs, only **(6 and 72)** and **(18 and 24)** have a GCF of 6 and an LCM of 72.

145. There are 33 twos and 28 fives in the prime factorization of this number. We can pair 28 twos with 28 fives to create 28 tens:
$$2^{33} \cdot 5^{28} = 2^5 \cdot 2^{28} \cdot 5^{28}$$
$$= 2^5 \cdot (2 \cdot 5)^{28}$$
$$= 2^5 \cdot 10^{28}.$$

So, this number is equal to 2^5 followed by 28 trailing zeros. Since $2^5 = 32$, the leftmost digit of the number is **3**.

The actual value of this number is 320,000,000,000,000,000,000,000,000,000.

146. Since $n!$ ends in 7 zeros, the greatest power of 10 that divides $n!$ is 10^7. Since $10^7 = (2 \cdot 5)^7 = 2^7 \cdot 5^7$, we know the prime factorization of $n!$ includes exactly seven 5's (and more than seven 2's).

So, we look for the smallest factorial that has exactly seven 5's in its prime factorization. Only multiples of 5 contribute 5's to the prime factorization. So, we only need to check values of n that are multiples of 5.

We begin with a guess of $n = 20$, and count the number of 5's in the prime factorization of $(n!)$.

If $n = 20$, the prime factorization of $n!$ includes four 5's.

$$20! = 20 \cdot \ \cdots \ \cdot 15 \cdot \ \cdots \ \cdot 10 \cdot \ \cdots \ \cdot 5 \cdot \ \cdots \ \cdot 1$$
$$\underset{5 \cdot 4}{} \qquad \underset{5 \cdot 3}{} \qquad \underset{5 \cdot 2}{} \qquad \underset{5 \cdot 1}{}$$

If $n = 25$, the prime factorization of $n!$ includes six 5's.

$$25! = 25 \cdot \ \cdots \ \cdot 20 \cdot \ \cdots \ \cdot 15 \cdot \ \cdots \ \cdot 10 \cdot \ \cdots \ \cdot 5 \cdot \ \cdots \ \cdot 1$$
$$\underset{5 \cdot 5}{} \qquad \underset{5 \cdot 4}{} \qquad \underset{5 \cdot 3}{} \qquad \underset{5 \cdot 2}{} \qquad \underset{5 \cdot 1}{}$$

If $n = 30$, the prime factorization of $n!$ includes seven 5's.

$$30! = 30 \cdot \ \cdots \ \cdot 25 \cdot \ \cdots \ \cdot 20 \cdot \ \cdots \ \cdot 15 \cdot \ \cdots \ \cdot 10 \cdot \ \cdots \ \cdot 5 \cdot \ \cdots \ \cdot 1$$
$$\underset{5 \cdot 6}{} \qquad \underset{5 \cdot 5}{} \qquad \underset{5 \cdot 4}{} \qquad \underset{5 \cdot 3}{} \qquad \underset{5 \cdot 2}{} \qquad \underset{5 \cdot 1}{}$$

These seven 5's can be paired with seven 2's to create seven 10's. Therefore, the smallest value of n for which $n!$ ends in 7 zeros is $n = \mathbf{30}$.

147. We write the prime factorization of $(10!)$:
$$10! = 10 \cdot 9 \cdot 8 \cdot 7 \cdot 6 \cdot 5 \cdot 4 \cdot 3 \cdot 2 \cdot 1$$
$$= (2 \cdot 5) \cdot (3^2) \cdot (2^3) \cdot 7 \cdot (2 \cdot 3) \cdot 5 \cdot (2^2) \cdot 3 \cdot 2$$
$$= 2^8 \cdot 3^4 \cdot 5^2 \cdot 7.$$

For the product of $10!$ and k to be a perfect square, the exponent of every prime in the prime factorization of $(10! \cdot k)$ must be even. The only prime whose exponent is not even in the prime factorization of $10!$ is 7.

So, if $k = 7$, then we have
$$10! \cdot k = 10! \cdot 7$$
$$= (2^8 \cdot 3^4 \cdot 5^2 \cdot 7) \cdot 7$$
$$= 2^8 \cdot 3^4 \cdot 5^2 \cdot 7^2.$$

Since every exponent in the prime factorization of this product is even, it is a perfect square. There is no smaller value of k that could make the exponent of 7 in this prime factorization even. So, **7** is the smallest positive value of k that makes the product of $10!$ and k a perfect square.

148. The LCM of $168 = 2^3 \cdot 3 \cdot 7$ and $980 = 2^2 \cdot 5 \cdot 7^2$ is
$$2^3 \cdot 3 \cdot 5 \cdot 7^2.$$

If Lizzie includes a third number and computes a different LCM, then either:

- the prime factorization of the new LCM includes a new prime factor, or

- the power of a prime factor in the new LCM is larger than its value in the original LCM.

The smallest prime factor that is not in $2^3 \cdot 3 \cdot 5 \cdot 7^2$ is 11. The smallest power of 2 larger than 2^3 is $2^4 = 16$. The smallest power of 3 larger than 3 is $3^2 = 9$. The smallest power of 5 larger than 5 is $5^2 = 25$. The smallest power of 7 larger than 7^2 is $7^3 = 343$.

Among the possibilities above, the smallest number that changes the prime factorization of the LCM is $3^2 = 9$. So, the smallest possible value of the third number is **9**.

— or —

The LCM of $168 = 2^3 \cdot 3 \cdot 7$ and $980 = 2^2 \cdot 5 \cdot 7^2$ is
$$2^3 \cdot 3 \cdot 5 \cdot 7^2.$$

Each integer from 1 to 8 is a factor of $2^3 \cdot 3 \cdot 5 \cdot 7^2$, as shown below.

$\underline{1}$: $\quad 2^3 \cdot 3 \cdot 5 \cdot 7^2 = 1 \cdot (2^3 \cdot 3 \cdot 5 \cdot 7^2)$
$\underline{2}$: $\quad 2^3 \cdot 3 \cdot 5 \cdot 7^2 = 2 \cdot (2^2 \cdot 3 \cdot 5 \cdot 7^2)$
$\underline{3}$: $\quad 2^3 \cdot 3 \cdot 5 \cdot 7^2 = 3 \cdot (2^3 \cdot 5 \cdot 7^2)$
$\underline{4}$: $\quad 2^3 \cdot 3 \cdot 5 \cdot 7^2 = 4 \cdot (2 \cdot 3 \cdot 5 \cdot 7^2)$
$\underline{5}$: $\quad 2^3 \cdot 3 \cdot 5 \cdot 7^2 = 5 \cdot (2^3 \cdot 3 \cdot 7^2)$
$\underline{6}$: $\quad 2^3 \cdot 3 \cdot 5 \cdot 7^2 = 6 \cdot (2^2 \cdot 5 \cdot 7^2)$
$\underline{7}$: $\quad 2^3 \cdot 3 \cdot 5 \cdot 7^2 = 7 \cdot (2^3 \cdot 3 \cdot 5 \cdot 7)$
$\underline{8}$: $\quad 2^3 \cdot 3 \cdot 5 \cdot 7^2 = 8 \cdot (3 \cdot 5 \cdot 7^2)$

However, $9 = 3^2$ is *not* a factor $2^3 \cdot 3 \cdot 5 \cdot 7^2$. So, the LCM of 9, 168, and 980 is different than the LCM of 168 and 980.

So, the smallest integer Lizzie could have included is **9**.

The LCM of 9, 168, and 980 is $2^3 \cdot 3^2 \cdot 5 \cdot 7^2$.

149. If x and y share any prime factors, we can always make one of x or y smaller so that they still have LCM 120.

For example, the numbers $x = 2^3 \cdot 5$ and $y = 2^2 \cdot 3$ have LCM 120, and they share 2^2 in their prime factorizations. Removing 2^2 from $y = 2^2 \cdot 3$ gives $y = 3$, and the LCM of $x = 2^3 \cdot 5$ and $y = 3$ is still 120.

So, we seek a pair of integers x and y that are relatively prime. The LCM of two numbers that are relatively prime is equal to the product of the two numbers. Therefore, $x \cdot y = 120$.

We consider the factor pairs of 120, crossing out pairs that are not relatively prime:

$$\underline{120}$$
$$1 \cdot 120 \qquad 5 \cdot 24$$
$$\cancel{2 \cdot 60} \qquad \cancel{6 \cdot 20}$$
$$3 \cdot 40 \qquad 8 \cdot 15$$
$$\cancel{4 \cdot 30} \qquad \cancel{10 \cdot 12}$$

The smallest sum we can get from a pair of relatively prime factors in this list is $8 + 15 = 23$. The LCM of 8 and 15 is 120, so the smallest possible value of $x + y$ is **23**.

150. Each factor pair of 48 represents possible dimensions for Alex's rectangle. For example, since $2 \cdot 24 = 48$, Alex could have a 2-inch-by-24-inch rectangle. Similarly, each factor pair of 72 represents possible dimensions for Grogg's rectangle.

Alex and Grogg are able to join their rectangles to create a larger rectangle, so their rectangles are joined along sides of the same length. So, the length of the side that Alex and Grogg join their rectangles along must be a common factor of 48 and 72.

The common factors of 48 and 72 are the factors of their GCF, 24. The factors of 24 are 1, 2, 3, 4, 6, 8, 12, and 24. These are the possible side lengths that Alex and Grogg could join their rectangles along.

This gives the following possibilities:

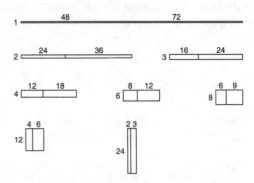

The rectangle with the smallest perimeter is the one that whose sides are closest in length. This is the 12-by-10 rectangle formed by joining the 12-by-4 and 12-by-6 rectangles. It has perimeter $(12 + 4 + 6) \cdot 2 = 22 \cdot 2 = $ **44 in**.

151. We write the prime factorization of each number.

12	14	15	20	30
$2^2 \cdot 3$	$2 \cdot 7$	$3 \cdot 5$	$2^2 \cdot 5$	$2 \cdot 3 \cdot 5$

We notice that every number has a 2 in its prime factorization except for 15. So, in order for the GCF of all three circled numbers to be 1, we must circle 15.

Then, since the GCF of any two circled numbers must be greater than 1, each of the other two circled numbers must share a prime factor with 15. We can eliminate 14, since it shares no prime factors with 15.

That leaves $12 = 2^2 \cdot 3$, $20 = 2^2 \cdot 5$, and $30 = 2 \cdot 3 \cdot 5$. If we circle 30, then circling either 12 or 20 will result in the GCF of all three numbers being greater than 1. So, we circle 12 and 20, as shown below.

$\left(\!\begin{array}{c}12\\2^2 \cdot 3\end{array}\!\right) \quad \begin{array}{c}14\\2 \cdot 7\end{array} \quad \left(\!\begin{array}{c}15\\3 \cdot 5\end{array}\!\right) \quad \left(\!\begin{array}{c}20\\2^2 \cdot 5\end{array}\!\right) \quad \begin{array}{c}30\\2 \cdot 3 \cdot 5\end{array}$

We check that the GCF of each pair of numbers is greater than 1, and the GCF of all three numbers is 1:
GCF(12, 15) = 3,
GCF(12, 20) = 4,
GCF(15, 20) = 5,
GCF(12, 15, 20) = 1. ✓

152. We consider the times when each pair of bells rings together.

Since LCM(4, 6) = 12, the 4-hour and 6-hour bells ring together every 12 hours. So, these bells ring together at the following hours:

12, 24, 36, 48, 60, 72, 84, 96.

Since LCM(4, 10) = 20, the 4-hour and 10-hour bells ring together every 20 hours. So, these bells ring together at the following hours:

20, 40, 60, 80, 100.

Since LCM(6, 10) = 30, the 6-hour and 10-hour bells ring together every 30 hours. So, these bells ring together at the following hours:

30, 60, 90.

We notice that 60 appears in all three lists. This is when all three bells ring together. Since we are counting the times at which *exactly two* bells ring, we count all of the times in the lists above except for 60 hours.

There are 7 times at which only the 4-hour and 6-hour bells ring together, 4 times at which only the 4-hour and 10-hour bells ring together, and 2 times at which only the 6-hour and 10-hour bells ring together.

So, exactly two bells will ring $7+4+2=$ **13** times in the next 100 hours.

153. We consider how the value of each coin balances around the average.

The gold coins are worth $50 each. So, each gold coin is $50-30=20$ dollars *above* the average.

The silver coins are worth $12 each. So, each silver coin is $30-12=18$ dollars *below* the average.

For these amounts to balance, the total amount that all gold coins are above the average must equal the total amount that all silver coins are below the average.

$$
\begin{array}{cc}
\overbrace{\begin{array}{cccc} {\scriptstyle +20} & {\scriptstyle +20} & {\scriptstyle \cdots} & {\scriptstyle +20 \;\; +20} \end{array}}^{+?} & \overbrace{\begin{array}{cccc} {\scriptstyle -18} & {\scriptstyle -18} & {\scriptstyle \cdots} & {\scriptstyle -18 \;\; -18} \end{array}}^{-?} \\
50 \;\; 50 \cdots 50 \;\; 50 & 12 \;\; 12 \cdots 12 \;\; 12 \\
\textit{Gold coins} & \textit{Silver coins}
\end{array}
$$

The total of the gold coins is above the average by some multiple of 20. The total of the silver coins is below the average by some multiple of 18.

So, these totals can only be equal at a common multiple of 20 and 18. To use the smallest number of coins, we compute the LCM of 20 and 18, which is $2^2 \cdot 3^2 \cdot 5 = 180$.

To get $180 above the average, we must use $180 \div 20 = 9$ gold coins.

To get $180 below the average, we must use $180 \div 18 = 10$ silver coins.

So, the smallest number of coins that could be in Kraken's satchel is $9+10=$ **19**.

We check that 9 gold coins and 10 silver coins have an average value of $30.

9 gold coins are worth $9 \cdot 50 = 450$ dollars.
10 silver coins are worth $10 \cdot 12 = 120$ dollars.
All together, the 19 coins are worth $450+120=570$ dollars, and their average value is $\frac{570}{19}=30$ dollars. ✓

Factors & Multiples Chapter 5 Solutions

1. Adding 8 seventeenths to 7 seventeenths, we get $8 + 7 = 15$ seventeenths. So, $\frac{8}{17} + \frac{7}{17} = \frac{15}{17}$.

2. Subtracting 4 thirteenths from 6 thirteenths leaves $6 - 4 = 2$ thirteenths. So, $\frac{6}{13} - \frac{4}{13} = \frac{2}{13}$.

3. $\frac{3}{10} + \frac{9}{10} = \frac{12}{10} = 1\frac{2}{10} = 1\frac{1}{5} = \frac{6}{5}$.

4. $\frac{3}{4} - \frac{1}{4} = \frac{2}{4} = \frac{1}{2}$.

5. $4\frac{5}{9} + 3\frac{2}{9} = 7\frac{7}{9} = \frac{70}{9}$.

6. $7\frac{4}{5} + 6\frac{3}{5} = 13\frac{7}{5} = 14\frac{2}{5} = \frac{72}{5}$.

7. $11\frac{7}{8} - 2\frac{5}{8} = 9\frac{2}{8} = 9\frac{1}{4} = \frac{37}{4}$.

8. $16 - 4\frac{2}{7} = 15\frac{7}{7} - 4\frac{2}{7} = 11\frac{5}{7} = \frac{82}{7}$.

9. We notice that 54 is a multiple of 9, so we have
$$\frac{2}{9} \cdot 54 = \frac{2 \cdot 54}{9} = 2 \cdot \frac{54}{9} = 2 \cdot 6 = \mathbf{12}.$$

10. $\frac{5}{6} \cdot 11 = \frac{5 \cdot 11}{6} = \frac{55}{6} = 9\frac{1}{6}$.

11. $\frac{3}{8} \cdot 7 = \frac{3 \cdot 7}{8} = \frac{21}{8} = 2\frac{5}{8}$.

12. We notice that 28 is a multiple of 14, so we have
$$\frac{5}{28} \cdot 14 = \frac{5 \cdot 14}{28} = 5 \cdot \frac{14}{28} = 5 \cdot \frac{1}{2} = \frac{5}{2} = 2\frac{1}{2}.$$

13. We notice that 36 is a multiple of 12, so we have
$$\frac{7}{12} \cdot 36 = \frac{7 \cdot 36}{12} = 7 \cdot \frac{36}{12} = 7 \cdot 3 = \mathbf{21}.$$

14. We notice that 26 is a multiple of 13, so we have
$$26 \cdot \frac{6}{13} = \frac{26 \cdot 6}{13} = \frac{26}{13} \cdot 6 = 2 \cdot 6 = \mathbf{12}.$$

15. We notice that 9 and 15 have a common factor of 3, so we have
$$9 \cdot \frac{8}{15} = \frac{9 \cdot 8}{15} = \frac{9}{15} \cdot 8 = \frac{3}{5} \cdot 8 = \frac{3 \cdot 8}{5} = \frac{24}{5} = 4\frac{4}{5}.$$

16. $3\frac{1}{2} = \frac{7}{2}$, so $3\frac{1}{2} \cdot 7 = \frac{7}{2} \cdot 7 = \frac{7 \cdot 7}{2} = \frac{49}{2} = \mathbf{24\frac{1}{2}}$.

— *or* —

$3\frac{1}{2} = 3 + \frac{1}{2}$, so we distribute:
$$\begin{aligned}
3\tfrac{1}{2} \cdot 7 &= \left(3 + \tfrac{1}{2}\right) \cdot 7 \\
&= (3 \cdot 7) + \left(\tfrac{1}{2} \cdot 7\right) \\
&= 21 + \tfrac{7}{2} \\
&= 21 + 3\tfrac{1}{2} \\
&= \mathbf{24\tfrac{1}{2}} \\
&= \frac{49}{2}.
\end{aligned}$$

17. We begin by converting $\frac{3}{5}$ into fifteenths: $\frac{3}{5} = \frac{9}{15}$. Then, we add: $\frac{3}{5} + \frac{2}{15} = \frac{9}{15} + \frac{2}{15} = \frac{11}{15}$.

18. $\frac{5}{8} = \frac{10}{16}$, so $\frac{5}{8} - \frac{5}{16} = \frac{10}{16} - \frac{5}{16} = \frac{5}{16}$.

19. $\frac{1}{4} = \frac{3}{12}$, so $\frac{5}{12} + \frac{1}{4} = \frac{5}{12} + \frac{3}{12} = \frac{8}{12} = \frac{2}{3}$.

20. $\frac{1}{3} = \frac{3}{9}$, so $\frac{5}{9} - \frac{1}{3} = \frac{5}{9} - \frac{3}{9} = \frac{2}{9}$.

21. $\frac{5}{6} = \frac{15}{18}$, so $\frac{17}{18} - \frac{5}{6} = \frac{17}{18} - \frac{15}{18} = \frac{2}{18} = \frac{1}{9}$.

22. $\frac{1}{2} = \frac{7}{14}$, so $\frac{1}{2} + \frac{3}{14} = \frac{7}{14} + \frac{3}{14} = \frac{10}{14} = \frac{5}{7}$.

23. $\frac{2}{3} = \frac{14}{21}$, so $4\frac{2}{3} - 2\frac{5}{21} = 4\frac{14}{21} - 2\frac{5}{21} = 2\frac{9}{21} = 2\frac{3}{7} = \frac{17}{7}$.

24. $\frac{2}{7} = \frac{18}{63}$, so $12\frac{61}{63} + 5\frac{2}{7} = 12\frac{61}{63} + 5\frac{18}{63} = 17\frac{79}{63} = 18\frac{16}{63} = \frac{1,150}{63}$.

In the remaining problems in this chapter, when adding or subtracting fractions, you may choose any common multiple for the denominator. We will always use the LCM of the original denominators.

25. The LCM of 3 and 4 is 12. So, we convert both $\frac{1}{3}$ and $\frac{1}{4}$ into twelfths.

$$\overset{\cdot 4}{\frac{1}{3}} = \frac{4}{12} \qquad \overset{\cdot 3}{\frac{1}{4}} = \frac{3}{12}$$
$$\underset{\cdot 4}{} \qquad\qquad \underset{\cdot 3}{}$$

Then, we add: $\frac{1}{3} + \frac{1}{4} = \frac{4}{12} + \frac{3}{12} = \frac{7}{12}$.

26. The LCM of 4 and 14 is 28. So we convert both $\frac{3}{4}$ and $\frac{5}{14}$ into twenty-eighths.

$\frac{3}{4} = \frac{21}{28}$ and $\frac{5}{14} = \frac{10}{28}$, so $\frac{3}{4} - \frac{5}{14} = \frac{21}{28} - \frac{10}{28} = \frac{11}{28}$.

27. LCM(4, 5) = 20. So, $\frac{2}{5} - \frac{1}{4} = \frac{8}{20} - \frac{5}{20} = \frac{3}{20}$.

28. LCM(6, 10) = 30. So, $\frac{1}{6} + \frac{7}{10} = \frac{5}{30} + \frac{21}{30} = \frac{26}{30} = \frac{13}{15}$.

29. LCM(6, 9) = 18. So, $\frac{1}{6} + \frac{8}{9} = \frac{3}{18} + \frac{16}{18} = \frac{19}{18} = 1\frac{1}{18}$.

30. We rewrite each mixed number as a fraction.
$$3\frac{4}{9} = \frac{31}{9} \text{ and } 1\frac{3}{10} = \frac{13}{10}.$$

We then convert the fractions so that they have the same denominator. The LCM of 9 and 10 is 90.
$$\frac{31}{9} = \frac{310}{90} \text{ and } \frac{13}{10} = \frac{117}{90}.$$

So, $3\frac{4}{9} - 1\frac{3}{10} = \frac{31}{9} - \frac{13}{10} = \frac{310}{90} - \frac{117}{90} = \frac{193}{90} = 2\frac{13}{90}$.

— *or* —

We convert only the fractional part of the mixed numbers. The LCM of 9 and 10 is 90.
$$\frac{4}{9} = \frac{40}{90} \text{ and } \frac{3}{10} = \frac{27}{90}.$$

So, $3\frac{4}{9} - 1\frac{3}{10} = 3\frac{40}{90} - 1\frac{27}{90} = 2\frac{13}{90} = \frac{193}{90}$.

31. The difference between the distances run is

$$3\frac{7}{8} - 2\frac{5}{6} = 3\frac{21}{24} - 2\frac{20}{24} = 1\frac{1}{24} \text{ miles.}$$

So, Dave ran $1\frac{1}{24} = \frac{25}{24}$ miles farther than Frank.

32. After Inchie's first climb and slip, he is

$$4\frac{1}{3} - 1\frac{1}{2} = 4\frac{2}{6} - 1\frac{3}{6} = 3\frac{8}{6} - 1\frac{3}{6} = 2\frac{5}{6}$$

inches from where he started.

Then, after he climbs up another $3\frac{3}{8}$ inches, he is

$$2\frac{5}{6} + 3\frac{3}{8} = 2\frac{20}{24} + 3\frac{9}{24} = 5\frac{29}{24} = 6\frac{5}{24} = \frac{149}{24}$$

inches from where he started.

33. The perimeter of the rectangle is

$$3\frac{1}{4} + 3\frac{1}{3} + 3\frac{1}{4} + 3\frac{1}{3} = 3\frac{3}{12} + 3\frac{4}{12} + 3\frac{3}{12} + 3\frac{4}{12}$$
$$= 12\frac{14}{12}$$
$$= 13\frac{2}{12}$$
$$= 13\frac{1}{6} \text{ feet.}$$

34. The sum of the known side lengths is

$$2\frac{1}{2} + 2 + 1\frac{1}{5} + 2\frac{1}{4} = 2\frac{10}{20} + 2 + 1\frac{4}{20} + 2\frac{5}{20}$$
$$= 7\frac{19}{20} \text{ meters.}$$

We subtract the sum of the known side lengths from the total perimeter to find the length of the remaining side:

$$9\frac{9}{20} - 7\frac{19}{20} = 8\frac{29}{20} - 7\frac{19}{20} = 1\frac{10}{20} = 1\frac{1}{2} \text{ meters.}$$

35. Since $\frac{2}{5} + \frac{1}{3} = \frac{6}{15} + \frac{5}{15} = \frac{11}{15}$, we know that $\frac{11}{15}$ of Tina's marbles are red or blue.

So, the remaining $1 - \frac{11}{15} = \frac{4}{15}$ of her marbles are yellow.

36. Subtracting a number and then adding the same number is the same as doing nothing:

$$\frac{1}{3} \underbrace{- \frac{1}{5} + \frac{1}{5}}_{+0} \underbrace{- \frac{1}{7} + \frac{1}{7}}_{+0} \underbrace{- \frac{1}{9} + \frac{1}{9}}_{+0} \underbrace{- \frac{1}{11} + \frac{1}{11}}_{+0} \underbrace{- \frac{1}{13} + \frac{1}{13}}_{+0} - \frac{1}{15}.$$

So, the expression simplifies to $\frac{1}{3} - \frac{1}{15} = \frac{5}{15} - \frac{1}{15} = \frac{4}{15}$.

37. a. In March, April, and May, Rosa grew a total of

$$\frac{1}{2} + 2\frac{1}{4} + \frac{1}{3} = \frac{6}{12} + 2\frac{3}{12} + \frac{4}{12} = 2\frac{13}{12} = 3\frac{1}{12} \text{ cm.}$$

So, Rosa was $3\frac{1}{12}$ **cm** taller at the end of May than she was at the beginning of March.

b. In May and June, Rosa grew a total of

$$\frac{1}{3} + 1\frac{3}{4} = \frac{4}{12} + 1\frac{9}{12} = 1\frac{13}{12} = 2\frac{1}{12} \text{ cm.}$$

Since Rosa was $157\frac{1}{3}$ cm tall at the end of June, her height at the end of April was

$$157\frac{1}{3} - 2\frac{1}{12} = 157\frac{4}{12} - 2\frac{1}{12}$$
$$= 155\frac{3}{12}$$
$$= 155\frac{1}{4} \text{ cm.}$$

c. During these six months, Rosa grew a total of

$$1\frac{1}{3} + \frac{5}{6} + \frac{1}{2} + 2\frac{1}{4} + \frac{1}{3} + 1\frac{3}{4}$$
$$= 1\frac{4}{12} + \frac{10}{12} + \frac{6}{12} + 2\frac{3}{12} + \frac{4}{12} + 1\frac{9}{12}$$
$$= 4\frac{36}{12}$$
$$= 7 \text{ cm.}$$

So, her average monthly growth was $\frac{7}{6} = 1\frac{1}{6}$ **cm.**

38. We use h to represent the weight of the hexatoad. Then, the weight of the octopug is $h + 1\frac{1}{2}$ pounds.

Together, the hexatoad and octopug weigh $10\frac{1}{6}$ pounds, so we have

$$h + \left(h + 1\frac{1}{2}\right) = 10\frac{1}{6}.$$

Subtracting $1\frac{1}{2}$ from both sides gives

$$h + h = 10\frac{1}{6} - 1\frac{1}{2}$$
$$= 10\frac{1}{6} - 1\frac{3}{6}$$
$$= 8\frac{4}{6}$$
$$= 8\frac{2}{3}.$$

Then, since $4\frac{1}{3} + 4\frac{1}{3} = 8\frac{2}{3}$, we have $h = 4\frac{1}{3}$. So, the hexatoad weighs $4\frac{1}{3}$ **pounds.**

Check: If the hexatoad weighs $4\frac{1}{3}$ pounds, then the octopug weighs $4\frac{1}{3} + 1\frac{1}{2} = 4\frac{2}{6} + 1\frac{3}{6} = 5\frac{5}{6}$ pounds.

Together, they weigh $4\frac{1}{3} + 5\frac{5}{6} = 4\frac{2}{6} + 5\frac{5}{6} = 9\frac{7}{6} = 10\frac{1}{6}$ pounds. ✓

39. Following the order of operations, we first evaluate the expressions grouped in parentheses, then multiply.

$$2\left(1 - \frac{1}{2}\right) + 3\left(1 - \frac{2}{3}\right) + \cdots + 99\left(1 - \frac{98}{99}\right) + 100\left(1 - \frac{99}{100}\right)$$
$$= 2\left(\frac{1}{2}\right) + 3\left(\frac{1}{3}\right) + \cdots + 99\left(\frac{1}{99}\right) + 100\left(\frac{1}{100}\right)$$
$$= 1 + 1 + \cdots + 1 + 1.$$

Each term in this sum simplifies to 1. There are 99 whole numbers from 2 to 100 (including 2 and 100), so this is the sum of 99 ones. The sum is equal to **99.**

40. We note that $\frac{1}{3} = \frac{2}{6}$, so $\frac{1}{6} + \frac{1}{6} = \frac{1}{3}$. However, we are told that A is less than B, so A and B are not both 6.

To get the sum of $\frac{1}{3}$ with different unit fractions, one fraction must be less than $\frac{1}{6}$ and the other must be greater than $\frac{1}{6}$.

The only unit fractions greater than $\frac{1}{6}$ are $\frac{1}{5}, \frac{1}{4}, \frac{1}{3}$, and $\frac{1}{2}$.

$\frac{1}{3}$ and $\frac{1}{2}$ are each greater than or equal to $\frac{1}{3}$, so we cannot use either of these fractions in our sum.

We check $\frac{1}{5}$ and $\frac{1}{4}$.

- $\frac{1}{5}$: Since $\frac{1}{3} - \frac{1}{5} = \frac{5}{15} - \frac{3}{15} = \frac{2}{15}$, we have $\frac{1}{5} + \frac{2}{15} = \frac{1}{3}$, which is not the sum of two unit fractions. ✗

- $\frac{1}{4}$: Since $\frac{1}{3} - \frac{1}{4} = \frac{4}{12} - \frac{3}{12} = \frac{1}{12}$, we have $\frac{1}{4} + \frac{1}{12} = \frac{1}{3}$, which is the sum of two unit fractions. ✓

Since A<B, we have **A = 4** and **B = 12.**

41. $\frac{1}{4}+\frac{1}{8}=\frac{2}{8}+\frac{1}{8}=\frac{3}{8}$.

42. $\frac{1}{5}+\frac{1}{20}=\frac{4}{20}+\frac{1}{20}=\frac{5}{20}=\frac{1}{4}$.

43. $\frac{1}{3}+\frac{1}{4}+\frac{1}{12}=\frac{4}{12}+\frac{3}{12}+\frac{1}{12}=\frac{8}{12}=\frac{2}{3}$.

44. $\frac{1}{2}+\frac{1}{5}+\frac{1}{10}=\frac{5}{10}+\frac{2}{10}+\frac{1}{10}=\frac{8}{10}=\frac{4}{5}$.

45. $\frac{5}{12}-\frac{1}{3}=\frac{5}{12}-\frac{4}{12}=\frac{1}{12}$. So, $\frac{5}{12}=\frac{1}{3}+\frac{1}{12}$.

46. $\frac{3}{11}-\frac{1}{4}=\frac{12}{44}-\frac{11}{44}=\frac{1}{44}$. So, $\frac{3}{11}=\frac{1}{4}+\frac{1}{\mathbf{44}}$.

47. $\frac{7}{9}-\frac{1}{2}-\frac{1}{4}=\frac{28}{36}-\frac{18}{36}-\frac{9}{36}=\frac{1}{36}$. So, $\frac{7}{9}=\frac{1}{2}+\frac{1}{4}+\frac{1}{\mathbf{36}}$.

48. $\frac{5}{11}-\frac{1}{3}-\frac{1}{99}=\frac{45}{99}-\frac{33}{99}-\frac{1}{99}=\frac{11}{99}=\frac{1}{9}$. So, $\frac{5}{11}=\frac{1}{3}+\frac{1}{\mathbf{9}}+\frac{1}{\mathbf{99}}$.

49. $\frac{1}{2}$ is the largest unit fraction that is less than $\frac{3}{4}$.

Subtracting $\frac{1}{2}$ from $\frac{3}{4}$ gives us $\frac{3}{4}-\frac{1}{2}=\frac{3}{4}-\frac{2}{4}=\frac{1}{4}$.

So, $\frac{3}{4}=\frac{1}{\mathbf{2}}+\frac{1}{\mathbf{4}}$.

50. $\frac{1}{2}$ is the largest unit fraction that is less than $\frac{2}{3}$.

Subtracting $\frac{1}{2}$ from $\frac{2}{3}$ gives us $\frac{2}{3}-\frac{1}{2}=\frac{4}{6}-\frac{3}{6}=\frac{1}{6}$.

So, $\frac{2}{3}=\frac{1}{\mathbf{2}}+\frac{1}{\mathbf{6}}$.

51. $\frac{1}{4}$ is the largest unit fraction that is less than $\frac{5}{16}$.

Subtracting $\frac{1}{4}$ from $\frac{5}{16}$ gives us $\frac{5}{16}-\frac{1}{4}=\frac{5}{16}-\frac{4}{16}=\frac{1}{16}$.

So, $\frac{5}{16}=\frac{1}{\mathbf{4}}+\frac{1}{\mathbf{16}}$.

52. $\frac{1}{2}$ is the largest unit fraction that is less than $\frac{4}{5}$.

Subtracting $\frac{1}{2}$ from $\frac{4}{5}$ gives us $\frac{4}{5}-\frac{1}{2}=\frac{8}{10}-\frac{5}{10}=\frac{3}{10}$.

So, we have $\frac{4}{5}=\frac{1}{2}+\frac{3}{10}$.

Next, we write $\frac{3}{10}$ as the sum of distinct unit fractions.

$\frac{1}{4}$ is the largest unit fraction that is less than $\frac{3}{10}$.

Subtracting $\frac{1}{4}$ from $\frac{3}{10}$ gives us $\frac{3}{10}-\frac{1}{4}=\frac{6}{20}-\frac{5}{20}=\frac{1}{20}$.

So, we have $\frac{3}{10}=\frac{1}{4}+\frac{1}{20}$.

Therefore, $\frac{4}{5}=\frac{1}{2}+\frac{3}{10}$
$=\frac{1}{\mathbf{2}}+\frac{1}{\mathbf{4}}+\frac{1}{\mathbf{20}}$.

53. Subtracting $\frac{1}{4}$ from $\frac{3}{7}$ gives us $\frac{3}{7}-\frac{1}{4}=\frac{12}{28}-\frac{7}{28}=\frac{5}{28}$.

So, we have $\frac{3}{7}=\frac{1}{4}+\frac{5}{28}$.

Next, we write $\frac{5}{28}$ as the sum of distinct unit fractions.

Since $\frac{5}{30}<\frac{5}{28}<\frac{5}{25}$, we know $\frac{1}{6}<\frac{5}{28}<\frac{1}{5}$.

So, $\frac{1}{6}$ is the largest unit fraction that is less than $\frac{5}{28}$.

Subtracting $\frac{1}{6}$ from $\frac{5}{28}$ gives us $\frac{5}{28}-\frac{1}{6}=\frac{15}{84}-\frac{14}{84}=\frac{1}{84}$.

So, we have $\frac{5}{28}=\frac{1}{6}+\frac{1}{84}$.

Therefore, $\frac{3}{7}=\frac{1}{4}+\frac{1}{6}+\frac{1}{\mathbf{84}}$.

54. $\frac{1}{5}-\frac{1}{6}=\frac{6}{30}-\frac{5}{30}=\frac{1}{30}$. So, $\frac{1}{5}=\frac{1}{6}+\frac{1}{\mathbf{30}}$.

55. $\frac{1}{9}-\frac{1}{10}=\frac{10}{90}-\frac{9}{90}=\frac{1}{90}$. So, $\frac{1}{9}=\frac{1}{10}+\frac{1}{\mathbf{90}}$.

56. $\frac{1}{8}-\frac{1}{9}=\frac{9}{72}-\frac{8}{72}=\frac{1}{72}$. So, $\frac{1}{8}=\frac{1}{9}+\frac{1}{\mathbf{72}}$.

57. $\frac{1}{99}-\frac{1}{100}=\frac{100}{9,900}-\frac{99}{9,900}=\frac{1}{9,900}$. So, $\frac{1}{99}=\frac{1}{100}+\frac{1}{\mathbf{9,900}}$.

58. We wish to compute $\frac{1}{n}-\frac{1}{n+1}$, but these two fractions do not have the same denominator.

The product $n\cdot(n+1)$ is a common multiple of n and $n+1$. So, we write both $\frac{1}{n}$ and $\frac{1}{n+1}$ as equivalent fractions with denominator $n\cdot(n+1)$.

$$\overset{\cdot(n+1)}{\overbrace{\frac{1}{n}}} = \frac{n+1}{n\cdot(n+1)} \qquad \overset{\cdot n}{\overbrace{\frac{1}{n+1}}} = \frac{n}{n\cdot(n+1)}$$

Now we subtract:

$$\frac{1}{n}-\frac{1}{n+1}=\frac{n+1}{n\cdot(n+1)}-\frac{n}{n\cdot(n+1)}=\frac{n+1-n}{n\cdot(n+1)}=\frac{1}{n\cdot(n+1)}.$$

So, we have $\frac{1}{n}=\frac{1}{n+1}+\frac{1}{\mathbf{n\cdot(n+1)}}$.

We can write this denominator $n\cdot(n+1)$ without the multiplication dot as $\mathbf{n(n+1)}$. We could also distribute the n and write this denominator as $\mathbf{n^2+n}$.

We notice that the Egyptian fractions in the four previous problems all follow this pattern!

54. $\frac{1}{5}=\frac{1}{6}+\frac{1}{5\cdot6}=\frac{1}{6}+\frac{1}{30}$ **55.** $\frac{1}{9}=\frac{1}{10}+\frac{1}{9\cdot10}=\frac{1}{10}+\frac{1}{90}$

56. $\frac{1}{8}=\frac{1}{9}+\frac{1}{8\cdot9}=\frac{1}{9}+\frac{1}{72}$ **57.** $\frac{1}{99}=\frac{1}{100}+\frac{1}{99\cdot100}=\frac{1}{100}+\frac{1}{9,900}$

59. In Problem 54, we found $\frac{1}{5}=\frac{1}{6}+\frac{1}{30}$.
$\frac{2}{5}=2\cdot\frac{1}{5}$, so $\frac{2}{5}=2\cdot\left(\frac{1}{6}+\frac{1}{30}\right)$.

Since 2 is a factor of both 6 and 30, when we distribute the 2, we get a sum of two distinct unit fractions:

$$\frac{2}{5}=2\cdot\left(\frac{1}{6}+\frac{1}{30}\right)$$
$$=\frac{2}{6}+\frac{2}{30}$$
$$=\frac{1}{\mathbf{3}}+\frac{1}{\mathbf{15}}.$$

60. In Problem 55, we found $\frac{1}{9}=\frac{1}{10}+\frac{1}{90}$.
$\frac{2}{9}=2\cdot\frac{1}{9}$, so $\frac{2}{9}=2\cdot\left(\frac{1}{10}+\frac{1}{90}\right)$.

Since 2 is a factor of both 10 and 90, when we distribute the 2, we get a sum of two distinct unit fractions:

$$\frac{2}{9}=2\cdot\left(\frac{1}{10}+\frac{1}{90}\right)$$
$$=\frac{2}{10}+\frac{2}{90}$$
$$=\frac{1}{\mathbf{5}}+\frac{1}{\mathbf{45}}.$$

61. In Problem 56, we found $\frac{1}{8}=\frac{1}{9}+\frac{1}{72}$.
$\frac{3}{8}=3\cdot\frac{1}{8}$, so $\frac{3}{8}=3\cdot\left(\frac{1}{9}+\frac{1}{72}\right)$.

Since 3 is a factor of both 9 and 72, when we distribute the 3, we get a sum of two distinct unit fractions:

$$\frac{3}{8}=3\cdot\left(\frac{1}{9}+\frac{1}{72}\right)$$
$$=\frac{3}{9}+\frac{3}{72}$$
$$=\frac{1}{\mathbf{3}}+\frac{1}{\mathbf{24}}.$$

62. In Problem 57, we found $\frac{1}{99} = \frac{1}{100} + \frac{1}{9,900}$.

$\frac{10}{99} = 10 \cdot \frac{1}{99}$, so $\frac{10}{99} = 10 \cdot \left(\frac{1}{100} + \frac{1}{9,900}\right)$.

Since 10 is a factor of both 100 and 9,900, when we distribute the 10, we get a sum of two distinct unit fractions:

$$\frac{10}{99} = 10 \cdot \left(\frac{1}{100} + \frac{1}{9,900}\right)$$
$$= \frac{10}{100} + \frac{10}{9,900}$$
$$= \frac{1}{\mathbf{10}} + \frac{1}{\mathbf{990}}.$$

63. $\frac{1}{2}$ is the largest unit fraction that is less than $\frac{8}{15}$.

Subtracting $\frac{1}{2}$ from $\frac{8}{15}$ gives us $\frac{8}{15} - \frac{1}{2} = \frac{16}{30} - \frac{15}{30} = \frac{1}{30}$.

So, we have $\frac{8}{15} = \frac{1}{2} + \frac{1}{30}$.

We found one sum with $\frac{1}{2}$, so we try a sum with $\frac{1}{3}$.

Subtracting $\frac{1}{3}$ from $\frac{8}{15}$ gives us $\frac{8}{15} - \frac{1}{3} = \frac{8}{15} - \frac{5}{15} = \frac{3}{15} = \frac{1}{5}$.

So, we have $\frac{8}{15} = \frac{1}{3} + \frac{1}{5}$.

Therefore, we have $\frac{8}{15} = \frac{1}{\mathbf{2}} + \frac{1}{\mathbf{30}}$ and $\frac{8}{15} = \frac{1}{\mathbf{3}} + \frac{1}{\mathbf{5}}$.

— or —

Using our observation from Problem 58, we have

$$\frac{1}{15} = \frac{1}{16} + \frac{1}{15 \cdot 16} = \frac{1}{16} + \frac{1}{240}.$$

Then, similar to our approaches in Problems 59-62, we have $\frac{8}{15} = 8 \cdot \frac{1}{15}$, so

$$\frac{8}{15} = 8 \cdot \left(\frac{1}{16} + \frac{1}{240}\right)$$
$$= \frac{8}{16} + \frac{8}{240}$$
$$= \frac{1}{\mathbf{2}} + \frac{1}{\mathbf{30}}.$$

We then find the second distinct pair as shown in the first solution. Therefore, we have $\frac{8}{15} = \frac{1}{\mathbf{2}} + \frac{1}{\mathbf{30}}$ and $\frac{8}{15} = \frac{1}{\mathbf{3}} + \frac{1}{\mathbf{5}}$.

Since $\frac{1}{4} + \frac{1}{4} = \frac{1}{2}$ is less than $\frac{8}{15}$, there is no way to write $\frac{8}{15}$ as the sum of two unit fractions that are both $\frac{1}{4}$ or less. So, the two solutions above are the only two solutions.

FRACTIONS

Crossout Sum   74-75

For each Crossout Sum puzzle, you may have found the missing numbers in a different order to arrive at the same final solution.

64. The missing number in the bottom row is

$$\frac{2}{3} - \frac{1}{4} = \frac{8}{12} - \frac{3}{12} = \frac{5}{12}.$$

The missing number in the right column is

$$\frac{7}{12} - \frac{5}{12} = \frac{2}{12} = \frac{1}{6}.$$

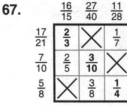

Then, we use either the row or column sum to find the remaining entry.

$$\frac{7}{24} - \frac{1}{6} = \frac{7}{24} - \frac{4}{24} = \frac{3}{24} = \frac{1}{8},$$
or $\frac{3}{8} - \frac{1}{4} = \frac{3}{8} - \frac{2}{8} = \frac{1}{8}.$

65. Step 1:   Step 2:   **Final:**

	$\frac{5}{8}$	$\frac{3}{4}$
$\frac{4}{5}$	$\frac{1}{4}$	
$\frac{23}{40}$	$\frac{3}{8}$	

	$\frac{5}{8}$	$\frac{3}{4}$
$\frac{4}{5}$	$\frac{1}{4}$	
$\frac{23}{40}$	$\frac{3}{8}$	$\frac{1}{5}$

	$\frac{5}{8}$	$\frac{3}{4}$
$\frac{4}{5}$	$\frac{1}{4}$	$\frac{11}{20}$
$\frac{23}{40}$	$\frac{3}{8}$	$\frac{1}{5}$

66. The missing number in the bottom row is

$$\frac{31}{45} - \frac{7}{15} = \frac{31}{45} - \frac{21}{45} = \frac{10}{45} = \frac{2}{9}.$$

The missing number in the middle column is

$$\frac{29}{21} - \frac{2}{3} = \frac{29}{21} - \frac{14}{21} = \frac{15}{21} = \frac{5}{7}.$$

The missing number in the right column is

$$\frac{7}{18} - \frac{1}{6} = \frac{7}{18} - \frac{3}{18} = \frac{4}{18} = \frac{2}{9}.$$

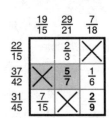

So, $\frac{2}{9}$ must be in the bottom-right square. Since there are exactly 2 numbers in each row and column, we X out the remaining squares in the bottom row and right column as shown.

Above, we found that the missing number in the middle column is $\frac{5}{7}$.

This number must be in the center square, and we X out the remaining square in the middle row.

We use the row or column sum to compute the remaining entry:

$$\frac{22}{15} - \frac{2}{3} = \frac{22}{15} - \frac{10}{15} = \frac{12}{15} = \frac{4}{5},$$
or $\frac{19}{15} - \frac{7}{15} = \frac{12}{15} = \frac{4}{5}.$

We use the strategies discussed previously to complete the remaining puzzles.

67.

	$\frac{16}{15}$	$\frac{27}{40}$	$\frac{11}{28}$
$\frac{17}{21}$	$\frac{2}{3}$	✕	$\frac{1}{7}$
$\frac{7}{10}$	$\frac{2}{5}$	$\frac{3}{10}$	✕
$\frac{5}{8}$	✕	$\frac{3}{8}$	$\frac{1}{4}$

68.

	$\frac{37}{28}$	$\frac{43}{30}$	$\frac{7}{9}$
$\frac{64}{63}$	$\frac{4}{7}$	✕	$\frac{4}{9}$
$\frac{27}{20}$	$\frac{3}{4}$	$\frac{3}{5}$	✕
$\frac{7}{6}$	✕	$\frac{5}{6}$	$\frac{1}{3}$

69.

	$\frac{31}{24}$	$\frac{4}{3}$	$\frac{31}{35}$
$\frac{49}{40}$	$\frac{5}{8}$	✕	$\frac{3}{5}$
$\frac{11}{14}$	✕	$\frac{1}{2}$	$\frac{2}{7}$
$\frac{3}{2}$	$\frac{2}{3}$	$\frac{5}{6}$	✕

70.

	$\frac{1}{2}$	$\frac{5}{6}$	$\frac{14}{15}$
$\frac{7}{12}$	$\frac{1}{4}$	✕	$\frac{1}{3}$
$\frac{19}{15}$	✕	$\frac{2}{3}$	$\frac{3}{5}$
$\frac{5}{12}$	$\frac{1}{4}$	$\frac{1}{6}$	✕

71.

	$\frac{79}{70}$	$\frac{25}{24}$	$\frac{13}{15}$
$\frac{5}{6}$	✕	$\frac{1}{6}$	$\frac{2}{3}$
$\frac{9}{10}$	$\frac{7}{10}$	✕	$\frac{1}{5}$
$\frac{73}{56}$	$\frac{3}{7}$	$\frac{7}{8}$	✕

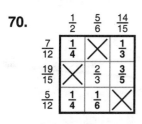

FRACTIONS

Fill-In Sums & Differences   76–77

72. We first look at $\frac{}{6}$. Since every fraction is less than one, 21 cannot be the numerator. Since every fraction is in simplest form, 4 cannot be the numerator. This leaves 1 as the numerator of $\frac{}{6}$.

$$\frac{1}{6}+\frac{}{}=\frac{5}{14}$$

The remaining numbers are 4 and 21. There is only one way to place these so that all fractions are less than one.

$$\frac{1}{6}+\frac{4}{21}=\frac{5}{14}$$

Check: $\frac{1}{6}+\frac{4}{21}=\frac{7}{42}+\frac{8}{42}=\frac{15}{42}=\frac{5}{14}$. ✓

73. Since every fraction is in simplest form, neither 2 nor 11 can be the numerator of $\frac{}{22}$. This leaves 1 as the numerator.

$$\frac{1}{22}+\frac{5}{}=\frac{}{}$$

The remaining numbers are 2 and 11. There is only one way to place these so that all fractions are less than one.

$$\frac{1}{22}+\frac{5}{11}=\frac{1}{2}$$

Check: $\frac{1}{22}+\frac{5}{11}=\frac{1}{22}+\frac{10}{22}=\frac{11}{22}=\frac{1}{2}$. ✓

74.

Step 1:	Step 2:	Final:
Numbers: *1*, 4, 5	Numbers: *1*, *4*, 5	Numbers: *1*, *4*, *5*

$$\frac{3}{}-\frac{}{15}=\frac{1}{3}\qquad \frac{3}{}-\frac{4}{15}=\frac{1}{3}\qquad \frac{3}{5}-\frac{4}{15}=\frac{1}{3}$$

Check: $\frac{3}{5}-\frac{4}{15}=\frac{9}{15}-\frac{4}{15}=\frac{5}{15}=\frac{1}{3}$. ✓

75. First, we place the 13. Since every fraction is less than one, 13 can only be the numerator of $\frac{}{18}$.

$$\frac{}{6}-\frac{13}{18}=\frac{}{9}$$

The remaining numbers are 1 and 5. We notice that $\frac{1}{6}$ is less than $\frac{13}{18}$, so we cannot subtract $\frac{13}{18}$ from $\frac{1}{6}$ and get a positive result.

$$\frac{\cancel{1}}{6}-\frac{13}{18}=\frac{}{9}$$

Therefore, the numerator of $\frac{}{6}$ is 5, leaving 1 as the numerator of $\frac{}{9}$.

$$\frac{5}{6}-\frac{13}{18}=\frac{1}{9}$$

Check: $\frac{5}{6}-\frac{13}{18}=\frac{15}{18}-\frac{13}{18}=\frac{2}{18}=\frac{1}{9}$. ✓

76. First, we look at $\frac{}{20}$. Since every fraction is in simplest form, 4 cannot be the numerator. So, either 3 or 17 is the numerator of $\frac{}{20}$.

If we choose 3 as the numerator of $\frac{}{20}$, then the remaining numbers are 4 and 17. Of these, only 4 can be the numerator of $\frac{}{5}$. However, $\frac{3}{20}$ is less than $\frac{4}{5}$, so we cannot subtract $\frac{4}{5}$ from $\frac{3}{20}$ and get a positive result.

$$\frac{\cancel{3}}{20}-\frac{\cancel{4}}{5}=\frac{1}{}$$

Therefore, the numerator of $\frac{}{20}$ is 17. The remaining numbers are 3 and 4. We try both possible placements.

$$\frac{17}{20}-\frac{}{5}=\frac{1}{}$$
$$\frac{17}{20}-\frac{3}{5}=\frac{1}{4}$$

$\frac{17}{20}-\frac{3}{5}=\frac{17}{20}-\frac{12}{20}=\frac{5}{20}=\frac{1}{4}$. ✓

$\frac{17}{20}-\frac{4}{5}=\frac{17}{20}-\frac{16}{20}=\frac{1}{20}$. ✗

77. Since every fraction is less than one, only 2 can be the numerator of $\frac{}{3}$.

$$\frac{}{}+\frac{}{}=\frac{2}{3}$$

Then, 24 is the largest number. Since every fraction is less than one, 24 is the denominator of one fraction.

$$\frac{}{24}+\frac{}{}=\frac{2}{3}$$

Of the remaining numbers, 3, 7, and 8, only 7 is *not* a factor of 24. So, 7 is the numerator of $\frac{}{24}$.

$$\frac{7}{24}+\frac{}{}=\frac{2}{3}$$

The remaining numbers are 3 and 8. There is only one way to place these so that all fractions are less than 1.

$$\frac{7}{24}+\frac{3}{8}=\frac{2}{3}$$

Addition is commutative, so you may have instead written $\frac{3}{8}+\frac{7}{24}=\frac{2}{3}$.

Check: $\frac{7}{24}+\frac{3}{8}=\frac{7}{24}+\frac{9}{24}=\frac{16}{24}=\frac{2}{3}$. ✓

78. Of the choices, only 1 or 9 could be the numerator of $\frac{}{10}$, and only 5 or 9 could be the denominator of $\frac{2}{}$.

$$\frac{2}{}+\frac{}{}=\frac{}{10}$$

However, since $\frac{1}{10}$ is less than both $\frac{2}{5}$ and $\frac{2}{9}$, the sum cannot equal $\frac{1}{10}$.

So, the numerator of $\frac{}{10}$ is 9. This leaves 5 as the denominator of $\frac{2}{}$.

$$\frac{2}{5}+\frac{}{}=\frac{9}{10}$$

The remaining numbers are 1 and 2. There is only one way to place these so that all fractions are less than 1.

$$\frac{2}{5}+\frac{1}{2}=\frac{9}{10}$$

Check: $\frac{2}{5}+\frac{1}{2}=\frac{4}{10}+\frac{5}{10}=\frac{9}{10}$. ✓

79. Of the choices, only 9 can be the numerator of $\frac{}{10}$.

$$\frac{9}{10}-\frac{1}{}=\frac{}{}$$

Then, 15 is the largest number and must be the denominator of one of the remaining fractions.

Neither of the two other remaining numbers, 5 and 6, can be the numerator of a fraction with denominator 15. So, 15 must be the denominator of $\frac{1}{}$.

$$\frac{9}{10}-\frac{1}{15}=\frac{}{}$$

The remaining numbers are 5 and 6. There is only one way to place these so that all fractions are less than 1.

$$\frac{9}{10}-\frac{1}{15}=\frac{5}{6}$$

Check: $\frac{9}{10}-\frac{1}{15}=\frac{27}{30}-\frac{2}{30}=\frac{25}{30}=\frac{5}{6}$. ✓

80. The denominator of the sum is 20, so the LCM of the two other denominators must be at least 20.

Among our number choices, only the pair (4, 5) has an LCM of at least 20. So, 4 and 5 are the denominators of the two empty fractions.

$$\frac{}{4}+\frac{}{5}=\frac{13}{20}$$

The remaining numbers are 1 and 2. There is only one way to place these so that all fractions are in simplest form.

$$\frac{1}{4}+\frac{2}{5}=\frac{13}{20}$$

Addition is commutative, so you may have instead written $\frac{2}{5}+\frac{1}{4}=\frac{13}{20}$.

Check: $\frac{1}{4}+\frac{2}{5}=\frac{5}{20}+\frac{8}{20}=\frac{13}{20}$. ✓

81. The denominator of the sum is 15, so the LCM of the two other denominators must be at least 15.

Among our number choices, only the pair (3, 5) has an LCM of at least 15. So, 3 and 5 are the denominators of the two empty fractions.

$$\frac{}{3}+\frac{}{5}=\frac{11}{15}$$

The two remaining numbers are 1 and 2. We try both possible placements.

$$\frac{1}{3}+\frac{2}{5}=\frac{11}{15}$$

$\frac{1}{3}+\frac{2}{5}=\frac{5}{15}+\frac{6}{15}=\frac{11}{15}$. ✓

$\frac{2}{3}+\frac{1}{5}=\frac{10}{15}+\frac{3}{15}=\frac{13}{15}$. ✗

Addition is commutative, so you may have instead written $\frac{2}{5}+\frac{1}{3}=\frac{11}{15}$.

82. Since all fractions are less than 1, 2 cannot be the denominator of any of these fractions. So, 2 is the missing numerator.

$$\frac{2}{}-\frac{4}{}=\frac{2}{}$$

The remaining numbers are 3, 7, and 21. The smallest remaining number, 3, can only be the denominator of one of the $\frac{2}{}$'s.

The denominator of the other $\frac{2}{}$ is 7 or 21.

Since $\frac{2}{3}$ is larger than both $\frac{2}{7}$ and $\frac{2}{21}$, we place the 3 as shown.

$$\frac{2}{3}-\frac{4}{}=\frac{2}{}$$

The remaining numbers are 7 and 21. We try both possible placements.

$$\frac{2}{3}-\frac{4}{7}=\frac{2}{21}$$

$\frac{2}{3}-\frac{4}{7}=\frac{14}{21}-\frac{12}{21}=\frac{2}{21}$. ✓

$\frac{2}{3}-\frac{4}{21}=\frac{14}{21}-\frac{4}{21}=\frac{10}{21}$. ✗

83. The largest number, 42, must be the denominator of one fraction.

$$\frac{}{42}+\frac{}{28}=\frac{}{12}$$

The second-largest number, 28, cannot be the numerator of $\frac{}{42}$, so it must be the denominator of another fraction.

The remaining numbers are 5, 7, and 13. Since 7 is a factor of both 42 and 28, we can only place 7 in the numerator of $\frac{}{12}$.

$$\frac{}{42}+\frac{}{28}=\frac{7}{12}$$

The remaining numbers are 5 and 13. We try both possible placements.

$$\frac{5}{42}+\frac{13}{28}=\frac{7}{12}$$

$\frac{5}{42}+\frac{13}{28}=\frac{10}{84}+\frac{39}{84}=\frac{49}{84}=\frac{7}{12}$. ✓

$\frac{13}{42}+\frac{5}{28}=\frac{26}{84}+\frac{15}{84}=\frac{41}{84}$. ✗

Addition is commutative, so you may have instead written $\frac{13}{28}+\frac{5}{42}=\frac{7}{12}$.

84. Since $\frac{4}{15}$ is less than $\frac{3}{5}=\frac{9}{15}$, the operation must be subtraction:

$$\frac{3}{5}\boxminus\frac{1}{3}=\frac{9}{15}-\frac{5}{15}=\frac{4}{15}. ✓$$

85. Since $\frac{11}{45}$ is greater than $\frac{1}{9}=\frac{5}{45}$, the operation must be addition:

$$\frac{1}{9}\boxplus\frac{2}{15}=\frac{5}{45}+\frac{6}{15}=\frac{11}{45}. ✓$$

86. We rewrite the fractions so that they all have the same denominator.

$$\frac{7}{8}\,\square\,\frac{1}{2}\,\square\,\frac{1}{8}=\frac{1}{2}\;\longrightarrow\;\frac{7}{8}\,\square\,\frac{4}{8}\,\square\,\frac{1}{8}=\frac{4}{8}$$

Now that all fractions are written as eighths, we focus on the numerators. We have $7-4+1=4$, so $\frac{7}{8}-\frac{4}{8}+\frac{1}{8}=\frac{4}{8}$.

$$\frac{7}{8}\boxminus\frac{1}{2}\boxplus\frac{1}{8}=\frac{1}{2}.$$

— *or* —

We start by computing the sum of the first two fractions:

$$\frac{7}{8}+\frac{1}{2}=\frac{7}{8}+\frac{4}{8}=\frac{11}{8}.$$

The final result of this problem is $\frac{1}{2}$, which is less than $\frac{11}{8}$. So, if the first operation is addition, then the second operation is subtraction.

In that case, the final result is $\frac{11}{8}-\frac{1}{8}=\frac{10}{8}=\frac{5}{4}$, not $\frac{1}{2}$. ✗

So, the first operation is subtraction: $\frac{7}{8}-\frac{1}{2}=\frac{7}{8}-\frac{4}{8}=\frac{3}{8}$.

The final result of this problem is $\frac{1}{2}$, which is greater than $\frac{3}{8}$. So, the second operation is addition:

$$\frac{7}{8}\boxminus\frac{1}{2}\boxplus\frac{1}{8}=\frac{7}{8}-\frac{4}{8}+\frac{1}{8}=\frac{4}{8}=\frac{1}{2}. ✓$$

We can use the approaches discussed in the previous problem to find the answers to the following problems.

87. $\frac{5}{12}\boxplus\frac{1}{3}\boxminus\frac{1}{2}=\frac{5}{12}+\frac{4}{12}-\frac{6}{12}=\frac{3}{12}=\frac{1}{4}.$ ✓

88. $\frac{1}{2}\boxplus\frac{1}{4}\boxplus\frac{3}{5}=\frac{10}{20}+\frac{5}{20}+\frac{12}{20}=\frac{27}{20}.$ ✓

89. $\frac{5}{6}\boxminus\frac{1}{4}\boxplus\frac{2}{5}=\frac{50}{60}-\frac{15}{60}+\frac{24}{60}=\frac{59}{60}.$ ✓

90. $\frac{13}{18}\boxminus\frac{1}{2}\boxplus\frac{1}{9}=\frac{13}{18}-\frac{9}{18}+\frac{2}{18}=\frac{6}{18}=\frac{1}{3}.$ ✓

91. $\frac{11}{20}\boxplus\frac{19}{24}\boxminus\frac{4}{5}=\frac{66}{120}+\frac{95}{120}-\frac{96}{120}=\frac{65}{120}=\frac{13}{24}.$ ✓

92. We locate $\frac{1}{2}$ on the number line by splitting the number line between 0 and 1 into two equal pieces.

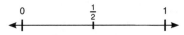

To find $\frac{1}{5}$ of $\frac{1}{2}$, we split each half into 5 equal pieces. Then, there are $2 \cdot 5 = 10$ equal pieces between 0 and 1. Each piece has length $\frac{1}{10}$.

Therefore, $\frac{1}{5}$ of $\frac{1}{2}$ is $\frac{1}{10}$.

93. $\frac{1}{3}$ of $\frac{1}{4}$ is $\frac{1}{12}$.

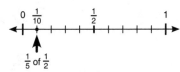

94. $\frac{1}{2}$ of $\frac{1}{3}$ is $\frac{1}{6}$.

95. $\frac{1}{4}$ of $\frac{1}{2}$ is $\frac{1}{8}$.

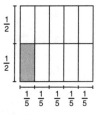

96. We split a unit square into fifths with four vertical lines and into halves with one vertical line.

The area of the small shaded rectangle is $\frac{1}{2} \cdot \frac{1}{5}$ square units.

Since the unit square is divided into 10 equal pieces, each is one tenth of the area of the large square. So, the area of the small shaded rectangle is $\frac{1}{10}$ square units.

Therefore, $\frac{1}{2} \cdot \frac{1}{5} = \frac{1}{10}$.

97. $\frac{1}{4} \cdot \frac{1}{3} = \frac{1}{12}$.

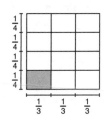

98. $\frac{1}{3} \cdot \frac{1}{6} = \frac{1}{18}$.

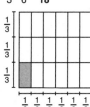

99. $\frac{1}{3} \cdot \frac{1}{4} = \frac{1}{12}$.

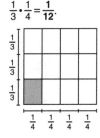

100. We know that *one* third of $\frac{1}{5}$ is $\frac{1}{15}$. *Two* thirds of $\frac{1}{5}$ is *two* times as much.

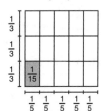 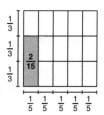

— *or* —

We have
$$\frac{2}{3} \cdot \frac{1}{5} = \left(2 \cdot \frac{1}{3}\right) \cdot \frac{1}{5}$$
$$= 2 \cdot \left(\frac{1}{3} \cdot \frac{1}{5}\right)$$
$$= 2 \cdot \frac{1}{15}$$
$$= \frac{2}{15}.$$

101.
$$\frac{9}{10} \cdot \frac{1}{10} = \left(9 \cdot \frac{1}{10}\right) \cdot \frac{1}{10}$$
$$= 9 \cdot \left(\frac{1}{10} \cdot \frac{1}{10}\right)$$
$$= 9 \cdot \frac{1}{100}$$
$$= \frac{9}{100}.$$

102.
$$\frac{1}{2} \cdot \frac{7}{11} = \frac{1}{2} \cdot \left(\frac{1}{11} \cdot 7\right)$$
$$= \left(\frac{1}{2} \cdot \frac{1}{11}\right) \cdot 7$$
$$= \frac{1}{22} \cdot 7$$
$$= \frac{7}{22}.$$

103.
$$\frac{1}{7} \cdot \frac{3}{4} = \frac{1}{7} \cdot \left(\frac{1}{4} \cdot 3\right)$$
$$= \left(\frac{1}{7} \cdot \frac{1}{4}\right) \cdot 3$$
$$= \frac{1}{28} \cdot 3$$
$$= \frac{3}{28}.$$

104.
$$\frac{3}{5} \cdot \frac{1}{4} = \left(3 \cdot \frac{1}{5}\right) \cdot \frac{1}{4}$$
$$= 3 \cdot \left(\frac{1}{5} \cdot \frac{1}{4}\right)$$
$$= 3 \cdot \frac{1}{20}$$
$$= \frac{3}{20}.$$

105.
$$\frac{1}{9} \cdot \frac{2}{3} = \frac{1}{9} \cdot \left(\frac{1}{3} \cdot 2\right)$$
$$= \left(\frac{1}{9} \cdot \frac{1}{3}\right) \cdot 2$$
$$= \frac{1}{27} \cdot 2$$
$$= \frac{2}{27}.$$

106.
$$\frac{7}{8} \cdot \frac{1}{5} = \left(7 \cdot \frac{1}{8}\right) \cdot \frac{1}{5}$$
$$= 7 \cdot \left(\frac{1}{8} \cdot \frac{1}{5}\right)$$
$$= 7 \cdot \frac{1}{40}$$
$$= \frac{7}{40}.$$

107.
$$\frac{5}{6} \cdot \frac{1}{3} = \left(5 \cdot \frac{1}{6}\right) \cdot \frac{1}{3}$$
$$= 5 \cdot \left(\frac{1}{6} \cdot \frac{1}{3}\right)$$
$$= 5 \cdot \frac{1}{18}$$
$$= \frac{5}{18}.$$

108.
$$\frac{7}{10} \cdot \frac{3}{4} = \left(7 \cdot \frac{1}{10}\right) \cdot \left(3 \cdot \frac{1}{4}\right)$$
$$= (7 \cdot 3) \cdot \left(\frac{1}{10} \cdot \frac{1}{4}\right)$$
$$= 21 \cdot \frac{1}{40}$$
$$= \frac{21}{40}.$$
— *or* —
$$\frac{7}{10} \cdot \frac{3}{4} = \frac{7 \cdot 3}{10 \cdot 4} = \frac{21}{40}.$$

109. $\frac{2}{3} \cdot \frac{5}{7} = \frac{2 \cdot 5}{3 \cdot 7} = \frac{10}{21}$.

110. $\frac{2}{13} \cdot \frac{3}{5} = \frac{2 \cdot 3}{13 \cdot 5} = \frac{6}{65}$.

111. $\frac{2}{7} \cdot \frac{4}{7} = \frac{2 \cdot 4}{7 \cdot 7} = \frac{8}{49}$.

112. $\frac{8}{9} \cdot \frac{4}{5} = \frac{8 \cdot 4}{9 \cdot 5} = \frac{32}{45}$.

113. $\frac{5}{8} \cdot \frac{9}{11} = \frac{5 \cdot 9}{8 \cdot 11} = \frac{45}{88}$.

114. $\frac{3}{4} \cdot \frac{7}{11} = \frac{3 \cdot 7}{4 \cdot 11} = \frac{21}{44}$.

115. $\frac{5}{6} \cdot \frac{5}{9} = \frac{5 \cdot 5}{6 \cdot 9} = \frac{25}{54}$.

116. $\frac{2}{3} \cdot \frac{1}{2} = \frac{2 \cdot 1}{3 \cdot 2} = \frac{2}{6} = \frac{1}{3}$.

— *or* —

We begin by cancelling the 2's. Then, we multiply:

$$\frac{\cancel{2}^1}{3} \cdot \frac{1}{\cancel{2}_1} = \frac{1 \cdot 1}{3 \cdot 1} = \frac{1}{3}.$$

117. $\frac{5}{6} \cdot \frac{6}{11} = \frac{5 \cdot 6}{6 \cdot 11} = \frac{30}{66} = \frac{5}{11}$. — *or* — $\frac{5}{\cancel{6}} \cdot \frac{\cancel{6}^1}{11} = \frac{5 \cdot 1}{1 \cdot 11} = \frac{5}{11}$.

118. $\frac{2}{3} \cdot \frac{5}{8} = \frac{2 \cdot 5}{3 \cdot 8} = \frac{10}{24} = \frac{5}{12}$. — *or* — $\frac{\cancel{2}^1}{3} \cdot \frac{5}{\cancel{8}_4} = \frac{1 \cdot 5}{3 \cdot 4} = \frac{5}{12}$.

119. $\frac{4}{5} \cdot \frac{3}{14} = \frac{4 \cdot 3}{5 \cdot 14} = \frac{12}{70} = \frac{6}{35}$. — *or* — $\frac{\cancel{4}^2}{5} \cdot \frac{3}{\cancel{14}_7} = \frac{2 \cdot 3}{5 \cdot 7} = \frac{6}{35}$.

120. We first divide 3 and 9 by their greatest common factor, 3.

$\frac{\cancel{3}^1}{10} \cdot \frac{4}{\cancel{9}_3}$

We then divide 4 and 10 by their greatest common factor, 2.

$\frac{\cancel{3}^1}{\cancel{10}_5} \cdot \frac{\cancel{4}^2}{\cancel{9}_3}$

Now that we have cancelled as much as we can, we compute the product.

$\frac{\cancel{3}^1}{\cancel{10}_5} \cdot \frac{\cancel{4}^2}{\cancel{9}_3} = \frac{1 \cdot 2}{5 \cdot 3} = \frac{2}{15}$

121. Step 1:
$\frac{\cancel{9}^3}{10} \cdot \frac{8}{\cancel{15}_5}$

Step 2:
$\frac{\cancel{9}^3}{\cancel{10}_5} \cdot \frac{\cancel{8}^4}{\cancel{15}_5}$

Step 3:
$\frac{\cancel{9}^3}{\cancel{10}_5} \cdot \frac{\cancel{8}^4}{\cancel{15}_5} = \frac{3 \cdot 4}{5 \cdot 5} = \frac{12}{25}$

122. Step 1:
$\frac{\cancel{3}^1}{8} \cdot \frac{5}{\cancel{6}_2} \cdot \frac{4}{5}$

Step 2:
$\frac{\cancel{3}^1}{8} \cdot \frac{\cancel{5}^1}{\cancel{6}_2} \cdot \frac{4}{\cancel{5}_1}$

Step 3:
$\frac{\cancel{3}^1}{\cancel{8}} \cdot \frac{\cancel{5}^1}{\cancel{6}_2} \cdot \frac{\cancel{4}^1}{\cancel{5}_1}$

Step 4:
$\frac{\cancel{3}^1}{\cancel{8}_2} \cdot \frac{\cancel{5}^1}{\cancel{6}_2} \cdot \frac{\cancel{4}^1}{\cancel{5}_1} = \frac{1 \cdot 1 \cdot 1}{2 \cdot 2 \cdot 1} = \frac{1}{4}$.

123. We cancel as many factors as possible before multiplying.

Step 1:
$\frac{\cancel{2}^1}{5} \cdot \frac{7}{10} \cdot \frac{20}{3} \cdot \frac{5}{21} \cdot \frac{9}{\cancel{2}_1}$

Step 2:
$\frac{\cancel{2}^1}{\cancel{5}_1} \cdot \frac{7}{10} \cdot \frac{20}{3} \cdot \frac{\cancel{5}^1}{21} \cdot \frac{9}{\cancel{2}_1}$

Step 3:
$\frac{\cancel{2}^1}{\cancel{5}_1} \cdot \frac{7}{\cancel{10}_1} \cdot \frac{\cancel{20}^2}{3} \cdot \frac{\cancel{5}^1}{21} \cdot \frac{9}{\cancel{2}_1}$

Step 4:
$\frac{\cancel{2}^1}{\cancel{5}_1} \cdot \frac{\cancel{7}^1}{\cancel{10}_1} \cdot \frac{\cancel{20}^2}{3} \cdot \frac{\cancel{5}^1}{\cancel{21}_3} \cdot \frac{9}{\cancel{2}_1}$

Step 5:
$\frac{\cancel{2}^1}{\cancel{5}_1} \cdot \frac{\cancel{7}^1}{\cancel{10}_1} \cdot \frac{\cancel{20}^2}{\cancel{3}_1} \cdot \frac{\cancel{5}^1}{\cancel{21}_3} \cdot \frac{\cancel{9}^3}{\cancel{2}_1}$

Step 6:
$\frac{\cancel{2}^1}{\cancel{5}_1} \cdot \frac{\cancel{7}^1}{\cancel{10}_1} \cdot \frac{\cancel{20}^2}{\cancel{3}_1} \cdot \frac{\cancel{5}^1}{\cancel{21}_{3}}^1 \cdot \frac{\cancel{9}^3}{\cancel{2}_1}$

Step 7:
$\frac{\cancel{2}^1}{\cancel{5}_1} \cdot \frac{\cancel{7}^1}{\cancel{10}_1} \cdot \frac{\cancel{20}^2}{\cancel{3}_1} \cdot \frac{\cancel{5}^1}{\cancel{21}} \cdot \frac{\cancel{9}^3}{\cancel{2}_1} = \frac{1 \cdot 1 \cdot 2 \cdot 1 \cdot 1}{1 \cdot 1 \cdot 1 \cdot 1 \cdot 1} = \frac{2}{1} = \mathbf{2}.$

124. Since the numerator of the product is 5, the denominator of $\frac{5}{}$ must cancel the factor of 6 in the numerator of $\frac{6}{7}$. So, we write a factor of 6 in the denominator of $\frac{5}{}$.

$\frac{5}{6 \cdot ?} \cdot \frac{6}{7} = \frac{5}{14}$

Since the denominator of the product is $14 = 2 \cdot 7$, we need another factor of 2 in the denominator of $\frac{5}{}$.

$\frac{5}{6 \cdot 2} \cdot \frac{6}{7} = \frac{5}{14}$

Therefore, the missing denominator is $6 \cdot 2 = 12$.

$\frac{5}{12} \cdot \frac{6}{7} = \frac{5}{14}$

Check: $\frac{5}{\cancel{12}_2} \cdot \frac{\cancel{6}^1}{7} = \frac{5 \cdot 1}{2 \cdot 7} = \frac{5}{14}$. ✓

125. The 2 and 40 have a GCF of 2, so 2 and 40 cancel as shown.

$\frac{\cancel{2}^1}{27} \cdot \frac{40}{\cancel{40}_{20}} = \frac{1}{60}$

The denominator of the product is $60 = 3 \cdot 20$, and $27 = 3^3$. So, the numerator of $\frac{}{40}$ must cancel a factor of $9 = 3^2$ in the denominator of $\frac{2}{27}$. We write a factor of 9 in the numerator of $\frac{}{40}$.

$\frac{\cancel{2}^1}{27} \cdot \frac{9 \cdot ?}{\cancel{40}_{20}} = \frac{1}{60}$

Since the numerator of the product is 1, the numerator of $\frac{}{40}$ has no additional factors.

$\frac{\cancel{2}^1}{27} \cdot \frac{9}{\cancel{40}_{20}} = \frac{1}{60}$

Therefore, the missing numerator is 9.

Check: $\frac{\cancel{2}^1}{\cancel{27}_3} \cdot \frac{\cancel{9}^1}{\cancel{40}_{20}} = \frac{1 \cdot 1}{3 \cdot 20} = \frac{1}{60}$. ✓

126. The 12 and 36 cancel as shown.

$\frac{5}{\cancel{36}_3} \cdot \frac{\cancel{12}^1}{15} = \frac{1}{15}$

Since the numerator of the product is 1, the denominator of $\frac{12}{}$ must cancel the factor of 5 in the numerator of $\frac{5}{36}$. So, we write a factor of 5 in the denominator of $\frac{12}{}$.

$\frac{5}{\cancel{36}_3} \cdot \frac{\cancel{12}^1}{5 \cdot ?} = \frac{1}{15}$

Since the denominator of the product is $15 = 3 \cdot 5$, we need another factor of 5 in the denominator of $\frac{12}{}$.

$\frac{5}{\cancel{36}_3} \cdot \frac{\cancel{12}^1}{5 \cdot 5} = \frac{1}{15}$

Therefore, the missing denominator is $5 \cdot 5 = 25$.

$\frac{5}{\cancel{36}_3} \cdot \frac{\cancel{12}^1}{25} = \frac{1}{15}$

Check: $\frac{\cancel{5}^1}{\cancel{36}_3} \cdot \frac{\cancel{12}^1}{\cancel{25}_5} = \frac{1 \cdot 1}{3 \cdot 5} = \frac{1}{15}$. ✓

127. The 8 and 24 cancel as shown.

$\frac{}{\cancel{24}_3} \cdot \frac{\cancel{8}^1}{65} = \frac{1}{39}$

The denominator of the product is $39 = 3 \cdot 13$, and $65 = 5 \cdot 13$. So, the numerator of $\frac{}{24}$ must cancel the factor of 5 in the denominator of $\frac{8}{65}$. So, we write a factor of 5 in the numerator of $\frac{}{24}$.

$$\frac{5 \cdot ?}{\overset{1}{\cancel{24}}_{3}} \cdot \frac{\overset{1}{\cancel{8}}}{65} = \frac{1}{39}$$

Since the numerator of the product is 1, the numerator of $\frac{}{24}$ has no additional factors.

$$\frac{5}{\overset{1}{\cancel{24}}_{3}} \cdot \frac{\overset{1}{\cancel{8}}}{65} = \frac{1}{39}$$

Therefore, the missing numerator is 5.

Check: $\frac{\overset{1}{\cancel{5}}}{\underset{3}{\cancel{24}}} \cdot \frac{\overset{1}{\cancel{8}}}{\underset{13}{\cancel{65}}} = \frac{1 \cdot 1}{3 \cdot 13} = \frac{1}{39}.$ ✔

128. The denominator of the product is $64 = 2^6$. The other two denominators are $24 = 2^3 \cdot 3$ and $40 = 2^3 \cdot 5$.

Therefore, the numerator of $\frac{}{40}$ must cancel the factor of 3 in the denominator of $\frac{}{24}$. So, we write a factor of 3 in the numerator of $\frac{}{40}$.

Similarly, the numerator of $\frac{}{24}$ must cancel the factor of 5 in the denominator of $\frac{}{40}$. So, we write a factor of 5 in the numerator of $\frac{}{24}$.

$$\frac{5 \cdot ?}{24} \cdot \frac{3 \cdot ?}{40} = \frac{3}{64}$$

Then, since the numerator of the product is 3, one of the other numerators must have an additional factor of 3.

Every fraction is in simplest form. Since 3 is a factor of 24, the numerator of $\frac{}{24}$ cannot be a multiple of 3.

So, the numerator of $\frac{}{40}$ must have the additional factor of 3.

$$\frac{5 \cdot ?}{24} \cdot \frac{3 \cdot 3}{40} = \frac{3}{64}$$

Therefore, the missing numerators are 5 and 9.

$$\frac{5}{24} \cdot \frac{9}{40} = \frac{3}{64}$$

Check: $\frac{\overset{1}{\cancel{5}}}{\underset{8}{\cancel{24}}} \cdot \frac{\overset{3}{\cancel{9}}}{\underset{8}{\cancel{40}}} = \frac{1 \cdot 3}{8 \cdot 8} = \frac{3}{64}.$ ✔

129. The denominator of the product is 3, and 16 is not a multiple of 3.

Therefore, the numerator of $\frac{}{33}$ must cancel the factor of 16 in the denominator of $\frac{}{16}$. So, we write a factor of 16 in the numerator of $\frac{}{33}$.

Then, the numerator of $\frac{}{16}$ must cancel the factor of 11 in the denominator of $\frac{}{33}$. So, we write a factor of 11 in the numerator of $\frac{}{16}$.

$$\frac{11 \cdot ?}{16} \cdot \frac{16 \cdot ?}{33} = \frac{1}{3}$$

Since the numerator of the product is 1, neither numerator has any additional factors.

$$\frac{11}{16} \cdot \frac{16}{33} = \frac{1}{3}$$

Therefore, the missing numerators are 11 and 16.

Check: $\frac{\overset{1}{\cancel{11}}}{\underset{1}{\cancel{16}}} \cdot \frac{\overset{1}{\cancel{16}}}{\underset{3}{\cancel{33}}} = \frac{1 \cdot 1}{1 \cdot 3} = \frac{1}{3}.$ ✔

130. The 20 and 32 cancel as shown.

$$\frac{}{\overset{}{\cancel{32}}_{8}} \cdot \frac{\overset{5}{\cancel{20}}}{} = \frac{45}{56}$$

Since the numerator of the product is $45 = 9 \cdot 5$, we write a factor of 9 in the numerator of $\frac{}{32}$.

$$\frac{9 \cdot ?}{\overset{}{\cancel{32}}_{8}} \cdot \frac{\overset{5}{\cancel{20}}}{} = \frac{45}{56}$$

Since the denominator of the product is $56 = 8 \cdot 7$, we write a factor of 7 in the denominator of $\frac{20}{}$.

$$\frac{9 \cdot ?}{\overset{}{\cancel{32}}_{8}} \cdot \frac{\overset{5}{\cancel{20}}}{7 \cdot ?} = \frac{45}{56}$$

All fractions are less than one, so the denominator of $\frac{20}{}$ cannot be 7.

$$\frac{9 \cdot ?}{\overset{}{\cancel{32}}_{8}} \cdot \frac{\overset{5}{\cancel{20}}}{7 \cdot 3} = \frac{45}{56}$$

Including a factor of 3 gives us a denominator of $7 \cdot 3 = 21$, which is greater than 20.

We also include another factor of 3 in the numerator of $\frac{}{32}$ to cancel the factor of 3 in the denominator of $\frac{20}{}$.

$$\frac{9 \cdot 3}{\overset{}{\cancel{32}}_{8}} \cdot \frac{\overset{5}{\cancel{20}}}{7 \cdot 3} = \frac{45}{56}$$

Therefore, the missing numerator is $9 \cdot 3 = 27$, and the missing denominator is $7 \cdot 3 = 21$.

$$\frac{27}{\overset{}{\cancel{32}}_{8}} \cdot \frac{\overset{5}{\cancel{20}}}{21} = \frac{45}{56}$$

If we chose any integer larger than 3 to be the additional factor, then the numerator of $\frac{}{32}$ would be at least $9 \cdot 4 = 36$, giving us a fraction greater than 1.

Check: $\frac{\overset{9}{\cancel{27}}}{\underset{8}{\cancel{32}}} \cdot \frac{\overset{5}{\cancel{20}}}{\underset{7}{\cancel{21}}} = \frac{9 \cdot 5}{8 \cdot 7} = \frac{45}{56}.$ ✔

131. The 8 and 10 cancel as shown.

$$\frac{\overset{4}{\cancel{8}}}{\underset{5}{\cancel{10}}} \cdot \frac{}{} = \frac{12}{25}$$

Since the numerator of the product is $12 = 4 \cdot 3$, we write a factor of 3 in the numerator of $\frac{}{10}$.

$$\frac{\overset{4}{\cancel{8}}}{\underset{5}{\cancel{10}}} \cdot \frac{3 \cdot ?}{} = \frac{12}{25}$$

Since the denominator of the product is $25 = 5 \cdot 5$, we write a factor of 5 in the denominator of $\frac{8}{}$.

$$\frac{\overset{4}{\cancel{8}}}{5 \cdot ?} \cdot \frac{3 \cdot ?}{\underset{5}{\cancel{10}}} = \frac{12}{25}$$

All fractions are less than 1, so 5 cannot be the denominator of $\frac{8}{}$.

Also, since all fractions are in simplest form and 2 is a factor of 8, the denominator of $\frac{8}{}$ cannot be even. So, we cannot include an additional factor of 2 in this denominator.

Including a factor of 3 in this denominator gives us a large enough denominator ($5 \cdot 3 = 15$, which is greater than 8) and a fraction in simplest form.

$$\frac{\overset{4}{\cancel{8}}}{5 \cdot 3} \cdot \frac{3 \cdot ?}{\underset{5}{\cancel{10}}} = \frac{12}{25}$$

We also include another factor of 3 in the numerator of $\frac{}{10}$ to cancel the factor of 3 in the denominator of $\frac{8}{}$.

$$\frac{\overset{4}{\cancel{8}}}{5 \cdot 3} \cdot \frac{3 \cdot 3}{\underset{5}{\cancel{10}}} = \frac{12}{25}$$

Therefore, the missing numerator is $3 \cdot 3 = 9$, and the missing denominator is $5 \cdot 3 = 15$.

$$\frac{\overset{4}{\cancel{8}}}{15} \cdot \frac{9}{\underset{5}{\cancel{10}}} = \frac{12}{25}$$

If we chose any integer larger than 3 to be the additional factor, then the numerator of $\frac{}{10}$ would be at least $3 \cdot 4 = 12$, giving us a fraction greater than 1.

Check: $\frac{\overset{4}{\cancel{8}}}{\underset{5}{\cancel{15}}} \cdot \frac{\overset{3}{\cancel{9}}}{\underset{5}{\cancel{10}}} = \frac{4 \cdot 3}{5 \cdot 5} = \frac{12}{25}.$ ✔

132. $1\frac{3}{5}=\frac{8}{5}$, so $1\frac{3}{5}\cdot\frac{1}{2}=\frac{8}{5}\cdot\frac{1}{2}=\frac{\overset{4}{8}}{5}\cdot\frac{1}{\underset{1}{2}}=\frac{4}{5}$.

— *or* —

$$1\frac{3}{5}\cdot\frac{1}{2}=\left(1+\frac{3}{5}\right)\cdot\frac{1}{2}$$
$$=\left(1\cdot\frac{1}{2}\right)+\left(\frac{3}{5}\cdot\frac{1}{2}\right)$$
$$=\frac{1}{2}+\frac{3}{10}$$
$$=\frac{5}{10}+\frac{3}{10}$$
$$=\frac{8}{10}$$
$$=\frac{4}{5}.$$

133. $\frac{5}{8}\cdot1\frac{1}{4}=\frac{5}{8}\cdot\frac{5}{4}=\frac{25}{32}$.

134. $2\frac{2}{3}\cdot\frac{3}{4}=\frac{8}{3}\cdot\frac{3}{4}=\frac{\overset{2}{\cancel{8}}}{\underset{1}{\cancel{3}}}\cdot\frac{\overset{1}{\cancel{3}}}{\underset{1}{\cancel{4}}}=2$.

135. $\frac{4}{5}\cdot3\frac{1}{8}=\frac{4}{5}\cdot\frac{25}{8}=\frac{\overset{1}{\cancel{4}}}{5}\cdot\frac{\overset{5}{\cancel{25}}}{\underset{2}{\cancel{8}}}=\frac{5}{2}=2\frac{1}{2}$.

136. $\frac{1}{6}\cdot9\frac{6}{7}=\frac{1}{6}\cdot\frac{69}{7}=\frac{1}{\cancel{6}}\cdot\frac{\overset{23}{\cancel{69}}}{7}=\frac{23}{14}=1\frac{9}{14}$.

137. $2\frac{1}{7}\cdot4\frac{1}{5}=\frac{15}{7}\cdot\frac{21}{5}=\frac{\overset{3}{\cancel{15}}}{\cancel{7}}\cdot\frac{\overset{3}{\cancel{21}}}{\cancel{5}}=9$.

138. $1\frac{7}{9}\cdot2\frac{1}{10}=\frac{16}{9}\cdot\frac{21}{10}=\frac{\overset{8}{\cancel{16}}}{\underset{3}{\cancel{9}}}\cdot\frac{\overset{7}{\cancel{21}}}{\underset{5}{\cancel{10}}}=\frac{56}{15}=3\frac{11}{15}$.

139. $8\frac{9}{14}\cdot\frac{7}{8}=\frac{121}{14}\cdot\frac{7}{8}=\frac{121}{\underset{2}{\cancel{14}}}\cdot\frac{\cancel{7}}{8}=\frac{121}{16}=7\frac{9}{16}$.

— *or* —

We notice that the denominator of $\frac{7}{8}$ is a factor of the whole-number part of the mixed number $8\frac{9}{14}$, so distributing will give us a whole number plus a fraction:

$$8\frac{9}{14}\cdot\frac{7}{8}=\left(8+\frac{9}{14}\right)\cdot\frac{7}{8}$$
$$=\left(8\cdot\frac{7}{8}\right)+\left(\frac{9}{14}\cdot\frac{7}{8}\right)$$
$$=7+\left(\frac{9}{\underset{2}{\cancel{14}}}\cdot\frac{\cancel{7}}{8}\right)$$
$$=7+\frac{9}{16}$$
$$=7\frac{9}{16}$$
$$=\frac{121}{16}.$$

140. Since $\frac{3}{8}$ of $\frac{4}{7}$ is $\frac{3}{8}\cdot\frac{4}{7}=\frac{3}{14}$, we know that $\frac{3}{14}$ of the students in Erin's class have dogs.

141. The area of the square is
$$2\frac{1}{10}\cdot2\frac{1}{10}=\frac{21}{10}\cdot\frac{21}{10}=\frac{441}{100}=4\frac{41}{100}\text{ square meters.}$$

142. Multiplying any positive number by a fraction that is less than one gives a result that is less than the original number. So, multiplying a positive number (a) by a fraction that is less than one (b) gives a product that is less than a. Therefore, $ab<a$. We are told that $a<b$, so from least to greatest, we have **$ab<a<b$**.

143. After Tim read $\frac{1}{4}$ of his book on Wednesday, $1-\frac{1}{4}=\frac{3}{4}$ of the book was left unread.

On Thursday, Tim read $\frac{2}{3}$ of the unread $\frac{3}{4}$ of the book. So, he read $\frac{2}{3}\cdot\frac{3}{4}=\frac{1}{2}$ of the whole book on Thursday.

All together, he read $\frac{1}{4}+\frac{1}{2}=\frac{1}{4}+\frac{2}{4}=\frac{3}{4}$ of the book on Wednesday and Thursday.

This leaves $1-\frac{3}{4}=\frac{1}{4}$ of the book to be read on Friday.

144. After Amy gives $\frac{2}{5}$ of her pennies to Ben, Amy is left with $1-\frac{2}{5}=\frac{3}{5}$ of her original pennies.

Then, after giving Chandra $\frac{3}{4}$ of these remaining pennies, Amy is left with $1-\frac{3}{4}=\frac{1}{4}$ of the remaining pennies.

So, after sharing with Ben and Chandra, Amy has $\frac{1}{4}$ of $\frac{3}{5}$ of her original pennies.

$\frac{1}{4}\cdot\frac{3}{5}=\frac{3}{20}$, so Amy's 12 wishing pennies are $\frac{3}{20}$ of her original pennies.

Three twentieths of Amy's original pennies is 12, so *one* twentieth of the original pennies is $12\div3=4$ pennies.

One twentieth of Amy's original pennies is 4, so Amy started with $4\cdot20=$ **80** pennies.

— *or* —

We work backwards.

Amy has 12 pennies after giving $\frac{3}{4}$ of her remaining pennies to Chandra. So, Amy's 12 pennies are $1-\frac{3}{4}=\frac{1}{4}$ of the pennies Amy had before sharing with Chandra.

So, before sharing with Chandra, Amy had $12\cdot4=48$ pennies.

These 48 pennies are what Amy had left after giving $\frac{2}{5}$ of her original pennies to Ben.

So, Amy's 48 pennies are $1-\frac{2}{5}=\frac{3}{5}$ of the pennies Amy originally had.

Three fifths of the pennies is 48, so *one* fifth of the pennies is $48\div3=16$.

One fifth of the pennies is 16, so the total number of pennies is $16\cdot5=80$.

Therefore, Amy originally had **80** pennies.

<u>Check:</u> Amy starts with 80 pennies and gives $\frac{2}{5}\cdot80=32$ pennies to Ben. Amy has $80-32=48$ pennies remaining.

Amy gives $\frac{3}{4}\cdot48=36$ of her remaining pennies to Chandra, leaving Amy with $48-36=12$ pennies to make wishes in a fountain. ✓

145. Since all fractions are less than 1, only 25 can be the denominator of $\frac{21}{}$.

$\frac{21}{25}\cdot\frac{}{}=\frac{42}{55}$

The remaining numbers are 10 and 11. There is only one way to place these so that all fractions are less than 1.

$\frac{21}{25}\cdot\frac{10}{11}=\frac{42}{55}$

Check: $\frac{21}{\underset{5}{\cancel{25}}}\cdot\frac{\overset{2}{\cancel{10}}}{11}=\frac{21\cdot2}{5\cdot11}=\frac{42}{55}$. ✓

146. Since all fractions are in simplest form, 22 cannot be the denominator of $\frac{11}{\;\;}$. Since all fractions are less than 1, and 22 is larger than all of the other number choices, 22 is the denominator of the empty fraction.

$$11 \cdot \frac{}{22} = \frac{5}{8}$$

The remaining numbers are 12 and 15. There is only one way to place these so that all fractions are in simplest form.

$$\frac{11}{12} \cdot \frac{15}{22} = \frac{5}{8}$$

Check: $\frac{\overset{1}{\cancel{11}}}{\underset{4}{12}} \cdot \frac{\overset{5}{\cancel{15}}}{\underset{2}{\cancel{22}}} = \frac{1 \cdot 5}{4 \cdot 2} = \frac{5}{8}.$ ✔

147. Since 5 is a factor of both 10 and 25, the 5 must be the denominator of $\frac{4}{\;\;}$.

$$\frac{4}{5} \cdot \frac{}{10} = \frac{}{25}$$

The remaining numbers are 3 and 6. There is only one way to place these so that all fractions are in simplest form.

$$\frac{4}{5} \cdot \frac{3}{10} = \frac{6}{25}$$

Check: $\frac{\overset{2}{\cancel{4}}}{5} \cdot \frac{3}{\underset{5}{\cancel{10}}} = \frac{2 \cdot 3}{5 \cdot 5} = \frac{6}{25}.$ ✔

148. Since the numerator of the product is 2, the factor of 5 in the numerator of $\frac{5}{6}$ must cancel with the denominator of the empty fraction.

Among our choices, only 5 is a multiple of 5. So, 5 is the denominator of the empty fraction.

$$\frac{5}{6} \cdot \frac{}{5} = \frac{2}{}$$

The remaining numbers are 3 and 4. There is only one way to place these so that all fractions are in simplest form.

$$\frac{5}{6} \cdot \frac{4}{5} = \frac{2}{3}$$

Check: $\frac{\overset{1}{\cancel{5}}}{\underset{3}{\cancel{6}}} \cdot \frac{\overset{2}{\cancel{4}}}{\underset{1}{\cancel{5}}} = \frac{1 \cdot 2}{3 \cdot 1} = \frac{2}{3}.$ ✔

149.

Step 1:	Step 2:	Final:
Numbers: 3, 7, ~~14~~	Numbers: ~~3~~, 7, ~~14~~	Numbers: ~~3~~, ~~7~~, ~~14~~
$\frac{9}{14} \cdot \frac{}{12} = \frac{}{8}$	$\frac{9}{14} \cdot \frac{}{12} = \frac{3}{8}$	$\frac{9}{14} \cdot \frac{7}{12} = \frac{3}{8}$

Check: $\frac{\overset{3}{\cancel{9}}}{\underset{2}{\cancel{14}}} \cdot \frac{\overset{1}{\cancel{7}}}{\underset{4}{\cancel{12}}} = \frac{3 \cdot 1}{2 \cdot 4} = \frac{3}{8}.$ ✔

150. Since the numerator of the product is 3, the factor of 5 in the numerator of $\frac{5}{\;\;}$ must cancel with the denominator of the empty fraction. Among our choices, only 5 and 25 are multiples of 5.

However, we do not have any number choice smaller than 5. So, if we choose 5 for the denominator of the empty fraction, we cannot make the fraction less than 1.

$$\frac{}{5} \cdot \frac{5}{} = \frac{3}{}$$

Therefore, 25 is the denominator of the empty fraction.

$$\frac{}{25} \cdot \frac{5}{} = \frac{3}{}$$

The remaining numbers are 5, 8, and 24. 5 can only be placed in the denominator of $\frac{3}{\;\;}$.

$$\frac{}{25} \cdot \frac{5}{} = \frac{3}{5}$$

The 5 and 25 cancel as shown.

$$\frac{}{\underset{5}{\cancel{25}}} \cdot \frac{\overset{1}{\cancel{5}}}{} = \frac{3}{5}$$

Since the numerator of the product is 3, and the numerator of $\frac{5}{\;\;}$ cancels as shown, the numerator of $\frac{}{25}$ must be a multiple of 3. Between our remaining numbers 8 and 24, only 24 is a multiple of 3.

$$\frac{24}{\underset{5}{\cancel{25}}} \cdot \frac{\overset{1}{\cancel{5}}}{} = \frac{3}{5}$$

Finally, 8 fills the remaining blank.

$$\frac{24}{\underset{5}{\cancel{25}}} \cdot \frac{\overset{1}{\cancel{5}}}{8} = \frac{3}{5}$$

Check: $\frac{\overset{3}{\cancel{24}}}{\underset{5}{\cancel{25}}} \cdot \frac{\overset{1}{\cancel{5}}}{\underset{1}{\cancel{8}}} = \frac{3 \cdot 1}{5 \cdot 1} = \frac{3}{5}.$ ✔

151.

Step 1:	Step 2:	Step 3:
Numbers: 3, 9, 15, ~~22~~	Numbers: 3, 9, 15, ~~22~~	Numbers: 3, 9, ~~15~~, ~~22~~
$\frac{}{22} \cdot \frac{11}{} = \frac{}{10}$	$\frac{}{\underset{2}{\cancel{22}}} \cdot \frac{\overset{1}{\cancel{11}}}{} = \frac{}{10}$	$\frac{}{\underset{2}{\cancel{22}}} \cdot \frac{\overset{1}{\cancel{11}}}{15} = \frac{}{10}$

The remaining numbers are 3 and 9. We try both possibilities.

$$\frac{\overset{1}{\cancel{3}}}{\underset{2}{\cancel{22}}} \cdot \frac{\overset{1}{\cancel{11}}}{\underset{5}{\cancel{15}}} = \frac{1 \cdot 1}{2 \cdot 5} = \frac{1}{10}. ✗$$

$$\frac{\overset{3}{\cancel{9}}}{\underset{2}{\cancel{22}}} \cdot \frac{\overset{1}{\cancel{11}}}{\underset{5}{\cancel{15}}} = \frac{3 \cdot 1}{2 \cdot 5} = \frac{3}{10}. ✔$$

152. Since all fractions are in simplest form, only 7 can be the numerator of $\frac{}{18}$.

$$\frac{7}{18} \cdot \frac{}{} = \frac{1}{}$$

Since the numerator of the product is 1, the factor of 7 in the numerator of $\frac{7}{18}$ must cancel with the denominator of the empty fraction.

Among our choices, only 14 is a multiple of 7. So, 14 is the denominator of the empty fraction.

$$\frac{7}{18} \cdot \frac{}{14} = \frac{1}{}$$

The remaining numbers are 4 and 9. There is only one way to place these so that all fractions are in simplest form.

$$\frac{7}{18} \cdot \frac{9}{14} = \frac{1}{4}$$

Check: $\frac{\overset{1}{\cancel{7}}}{\underset{2}{\cancel{18}}} \cdot \frac{\overset{1}{\cancel{9}}}{\underset{2}{\cancel{14}}} = \frac{1 \cdot 1}{2 \cdot 2} = \frac{1}{4}.$ ✔

153. Among the number choices, only 4 can be the numerator of $\frac{}{9}$.

$$\frac{}{} \cdot \frac{}{} = \frac{4}{9}$$

21 is the largest number among the choices, so it must be the denominator of one of the empty fractions.

$$\frac{}{} \cdot \frac{}{21} = \frac{4}{9}$$

The remaining numbers are 10, 14, and 15. Among these choices, only 10 can be the numerator of $\frac{}{21}$.

$$\frac{}{} \cdot \frac{10}{21} = \frac{4}{9}$$

The remaining numbers are 14 and 15. There is only one way to place these so that all fractions are less than 1.

$$\frac{14}{15} \cdot \frac{10}{21} = \frac{4}{9}$$

Multiplication is commutative, so you may have instead written $\frac{10}{21} \cdot \frac{14}{15} = \frac{4}{9}$.

Check: $\frac{\overset{2}{\cancel{14}}}{\underset{3}{\cancel{15}}} \cdot \frac{\overset{2}{\cancel{10}}}{\underset{3}{\cancel{21}}} = \frac{2 \cdot 2}{3 \cdot 3} = \frac{4}{9}.$ ✔

154. Since the numerator of the product is 5, at least one of the numerators of the empty fractions must be a multiple of 5. Among our choices, only 5 is a multiple of 5.

$$\frac{5}{}\cdot\frac{}{}=\frac{5}{}$$

The smallest number choice, 2, must be the numerator of the remaining empty fraction.

$$\frac{5}{}\cdot\frac{2}{}=\frac{5}{}$$

Since the numerator of the product is 5, the factor of 2 in the numerator of $\frac{2}{}$ must cancel with the denominator of $\frac{5}{}$.

Among our choices, only 6 is a multiple of 2. So, 6 is the denominator of $\frac{5}{}$.

$$\frac{5}{6}\cdot\frac{2}{}=\frac{5}{}$$

The 2 and 6 cancel as shown.

$$\frac{5}{\overset{}{\underset{3}{6}}}\cdot\frac{\overset{1}{2}}{}=\frac{5}{}$$

The factor of 3 in the denominator of $\frac{5}{6}$ does not cancel. Therefore, the denominator of the product is also a multiple of 3.

Between our choices of 7 and 21, only 21 is a multiple of 3.

$$\frac{5}{\overset{}{\underset{3}{6}}}\cdot\frac{\overset{1}{2}}{}=\frac{5}{21}$$

Finally, 7 fills the remaining blank.

$$\frac{5}{\overset{}{\underset{3}{6}}}\cdot\frac{\overset{1}{2}}{7}=\frac{5}{21}$$

Multiplication is commutative, so you may have instead written $\frac{2}{7}\cdot\frac{5}{6}=\frac{5}{21}$.

Check: $\frac{5}{\overset{}{\underset{3}{6}}}\cdot\frac{\overset{1}{2}}{7}=\frac{5\cdot1}{3\cdot7}=\frac{5}{21}.$ ✓

155. Since 2 is the smallest number in the puzzle, it must be the numerator of one of the empty fractions.

$$\frac{2}{}\cdot\frac{}{}=\frac{3}{}$$

The remaining numbers are 3, 5, 9, and 10. Since the numerator of the product is 3, the factor of 2 in the numerator of $\frac{2}{}$ must cancel with the denominator of the empty fraction.

Among our choices, only 10 is a multiple of 2.

$$\frac{2}{}\cdot\frac{}{10}=\frac{3}{}$$

The 2 and 10 cancel as shown.

$$\frac{\overset{1}{2}}{}\cdot\frac{}{\underset{5}{10}}=\frac{3}{}$$

The factor of 5 in the denominator of $\frac{}{10}$ does not cancel. Therefore, the denominator of the product is also a multiple of 5.

Among our remaining choices, only 5 is a multiple of 5.

$$\frac{\overset{1}{2}}{}\cdot\frac{}{\underset{5}{10}}=\frac{3}{5}$$

The remaining 3 and 9 must fill the blank numerator and denominator. However we arrange 3 and 9 in these two blanks, their common factor of 3 will cancel.

So, to get a product with numerator 3 and denominator 5, we must place the 3 and 9 as shown.

$$\frac{\overset{1}{2}}{3}\cdot\frac{9}{\underset{5}{10}}=\frac{3}{5}$$

Multiplication is commutative, so you may have instead written $\frac{9}{10}\cdot\frac{2}{3}=\frac{3}{5}$.

Check: $\frac{\overset{1}{2}}{\underset{1}{3}}\cdot\frac{\overset{3}{9}}{\underset{5}{10}}=\frac{1\cdot3}{1\cdot5}=\frac{3}{5}.$ ✓

156. The numerator of the product is 1. So, the denominator of $\frac{5}{}$ must be a multiple of the numerator of the empty fraction.

12 is the only number choice that is a multiple of another choice (6). So, we place 6 and 12 as shown so that the denominator of $\frac{5}{}$ is a multiple of the numerator of the empty fraction.

$$\frac{5}{12}\cdot\frac{6}{}=\frac{1}{}$$

Then, the 6 and 12 cancel as shown.

$$\frac{5}{\underset{2}{12}}\cdot\frac{\overset{1}{6}}{}=\frac{1}{}$$

The second factor of 2 in the denominator of $\frac{5}{12}$ does not cancel. Therefore, the denominator of the product is a multiple of 2.

Between our choices of 10 and 25, only 10 is a multiple of 2.

$$\frac{5}{\underset{2}{12}}\cdot\frac{\overset{1}{6}}{}=\frac{1}{10}$$

Finally, 25 fills the remaining blank.

$$\frac{5}{\underset{2}{12}}\cdot\frac{\overset{1}{6}}{25}=\frac{1}{10}$$

Check: $\frac{\overset{1}{5}}{\underset{2}{12}}\cdot\frac{\overset{1}{6}}{\underset{5}{25}}=\frac{1\cdot1}{2\cdot5}=\frac{1}{10}.$ ✓

157. $\frac{1}{7}\cdot7=1$, so the reciprocal of $\frac{1}{7}$ is **7**.

158. $25\cdot\frac{1}{25}=1$, so the reciprocal of 25 is $\frac{1}{25}$.

159. $2+8=10$, and $10\cdot\frac{1}{10}=1$. So, the reciprocal of $2+8$ is $\frac{1}{10}$.

— *or* —

We see that $(2+8)\cdot\frac{1}{(2+8)}=1$. So, we can write the reciprocal of $2+8$ as $\frac{1}{(2+8)}$, which simplifies to $\frac{1}{10}$.

160. $9\cdot4=36$, and $36\cdot\frac{1}{36}=1$. So, the reciprocal of $9\cdot4$ is $\frac{1}{36}$.

— *or* —

We see that $(9\cdot4)\cdot\frac{1}{(9\cdot4)}=1$. So, we can write the reciprocal of $9\cdot4$ as $\frac{1}{9\cdot4}$, which simplifies to $\frac{1}{36}$.

161. We first simplify: $\frac{24}{6}=4$. Then, we have $4\cdot\frac{1}{4}=1$. So, the reciprocal of $\frac{24}{6}$ is $\frac{1}{4}$.

162. We first simplify: $\frac{11}{55}=\frac{1}{5}$. Then, we have $\frac{1}{5}\cdot5=1$. So, the reciprocal of $\frac{11}{55}$ is **5**.

163. $7\div\frac{1}{5}=7\cdot5=\textbf{35}$.

164. $5\div\frac{1}{6}=5\cdot6=\textbf{30}$.

165. $2\div\left(5\div\frac{1}{8}\right)=2\div(5\cdot8)=2\div40=\frac{2}{40}=\frac{1}{20}$.

166. $3\div\left(\frac{1}{4}\div\frac{1}{2}\right)=3\div\left(\frac{1}{4}\cdot2\right)=3\div\frac{2}{4}=3\div\frac{1}{2}=3\cdot2=\textbf{6}$.

167. Since dividing by a number is the same as multiplying by its reciprocal, $a\div\frac{1}{23}=17$ means the same as $a\cdot23=17$. If $a\cdot23=17$, then $a=17\div23=\frac{17}{23}$.

168. $\frac{2}{3}\cdot\frac{3}{2}=1$, so the reciprocal of $\frac{2}{3}$ is $\frac{3}{2}=1\frac{1}{2}$.

169. $\frac{3}{8} \cdot \frac{8}{3} = 1$, so the reciprocal of $\frac{3}{8}$ is $\frac{8}{3} = 2\frac{2}{3}$.

170. $2\frac{1}{2} = \frac{5}{2}$, and $\frac{5}{2} \cdot \frac{2}{5} = 1$. So, the reciprocal of $2\frac{1}{2}$ is $\frac{2}{5}$.

171. $5\frac{1}{3} = \frac{16}{3}$, and $\frac{16}{3} \cdot \frac{3}{16} = 1$. So, the reciprocal of $5\frac{1}{3}$ is $\frac{3}{16}$.

172. $2\frac{2}{5} = \frac{12}{5}$, and $\frac{12}{5} \cdot \frac{5}{12} = 1$. So, the reciprocal of $2\frac{2}{5}$ is $\frac{5}{12}$.

173. $4 \cdot \frac{5}{9} = \frac{20}{9}$, and $\frac{20}{9} \cdot \frac{9}{20} = 1$. So, the reciprocal of $4 \cdot \frac{5}{9}$ is $\frac{9}{20}$.

174. $\frac{4}{7} \cdot \frac{3}{5} = \frac{12}{35}$, and $\frac{12}{35} \cdot \frac{35}{12} = 1$.
So, the reciprocal of $\frac{4}{7} \cdot \frac{3}{5}$ is $\frac{35}{12} = 2\frac{11}{12}$.

175. $\frac{2}{5} + \frac{2}{9} = \frac{18}{45} + \frac{10}{45} = \frac{28}{45}$, and $\frac{28}{45} \cdot \frac{45}{28} = 1$.
So, the reciprocal of $\frac{2}{5} + \frac{2}{9}$ is $\frac{45}{28} = 1\frac{17}{28}$.

176. $\frac{1}{2} + \frac{1}{3} + \frac{1}{4} = \frac{6}{12} + \frac{4}{12} + \frac{3}{12} = \frac{13}{12}$.
Since $\frac{13}{12} \cdot \frac{12}{13} = 1$, the reciprocal of the sum of one half, one third, and one fourth is $\frac{12}{13}$.

FRACTIONS
Division
91

177. $9 \div \frac{3}{4} = 9 \cdot \frac{4}{3} = \frac{36}{3} = 12$.

178. $\frac{5}{8} \div \frac{7}{9} = \frac{5}{8} \cdot \frac{9}{7} = \frac{45}{56}$.

179. $\frac{1}{4} \div \frac{3}{8} = \frac{1}{4} \cdot \frac{8}{3} = \frac{1}{4} \cdot \frac{8}{3}^2 = \frac{2}{3}$.

180. $\frac{5}{6} \div \frac{4}{5} = \frac{5}{6} \cdot \frac{5}{4} = \frac{25}{24} = 1\frac{1}{24}$.

181. $1\frac{5}{8} = \frac{13}{8}$, so $1\frac{5}{8} \div \frac{2}{3} = \frac{13}{8} \cdot \frac{3}{2} = \frac{39}{16} = 2\frac{7}{16}$.

182. $\frac{8}{15} \div \frac{12}{25} = \frac{8}{15} \cdot \frac{25}{12} = \frac{8}{15} \cdot \frac{25}{12} = \frac{10}{9} = 1\frac{1}{9}$.

183. $\frac{21}{16} \div \frac{49}{50} = \frac{21}{16} \cdot \frac{50}{49} = \frac{21}{16} \cdot \frac{50}{49} = \frac{75}{56} = 1\frac{19}{56}$.

184. $2\frac{6}{11} = \frac{28}{11}$. So, $\frac{52}{55} \div 2\frac{6}{11} = \frac{52}{55} \div \frac{28}{11} = \frac{52}{55} \cdot \frac{11}{28} = \frac{52}{55} \cdot \frac{11}{28} = \frac{13}{35}$.

185. $\frac{6}{35} \div \left(\frac{3}{14} \div \frac{4}{5}\right) = \frac{6}{35} \div \left(\frac{3}{14} \cdot \frac{5}{4}\right) = \frac{6}{35} \div \frac{15}{56} = \frac{6}{35} \cdot \frac{56}{15} = \frac{6}{35} \cdot \frac{56}{15} = \frac{16}{25}$.

186. $\left(\frac{6}{35} \div \frac{3}{14}\right) \div \frac{4}{5} = \left(\frac{6}{35} \cdot \frac{14}{3}\right) \div \frac{4}{5} = \frac{6}{35} \cdot \frac{14}{3} \div \frac{4}{5} = \frac{4}{5} \div \frac{4}{5} = 1$.

FRACTIONS
Multiplication Tables
92-93

187. We look for the number in the top-left shaded square. Since $\frac{1}{4} \cdot \square = \frac{1}{10}$, we know $\frac{1}{10} \div \frac{1}{4} = \square$.
$\frac{1}{10} \div \frac{1}{4} = \frac{1}{10} \cdot 4 = \frac{4}{10} = \frac{2}{5}$, so the number in this square is $\frac{2}{5}$.

·	2/5	3/5	
1/4	1/10		1/5
		9/20	

Next, we consider the top-right shaded square. Since $\frac{1}{4} \cdot \square = \frac{1}{5}$, we know $\frac{1}{5} \div \frac{1}{4} = \square$.
$\frac{1}{5} \div \frac{1}{4} = \frac{1}{5} \cdot 4 = \frac{4}{5}$, so the number in this square is $\frac{4}{5}$.

·	2/5	3/5	4/5
1/4	1/10		1/5
		9/20	

Then, we consider the bottom-left shaded square. Since $\frac{3}{5} \cdot \square = \frac{9}{20}$, we know $\frac{9}{20} \div \frac{3}{5} = \square$.
$\frac{9}{20} \div \frac{3}{5} = \frac{9}{20} \cdot \frac{5}{3} = \frac{3}{4}$, so the number in this square is $\frac{3}{4}$.

·	2/5	3/5	4/5
1/4	1/10		1/5
3/4		9/20	

Finally, we complete the puzzle as shown with the following products:
$\frac{1}{4} \cdot \frac{3}{5} = \frac{3}{20}$,
$\frac{3}{4} \cdot \frac{2}{5} = \frac{3}{10}$,
$\frac{3}{4} \cdot \frac{4}{5} = \frac{3}{5}$.

·	2/5	3/5	4/5
1/4	1/10	3/20	1/5
3/4	3/10	9/20	3/5

188.

Step 1:
$\frac{1}{6} \div \frac{5}{6} = \frac{1}{6} \cdot \frac{6}{5} = \frac{1}{5}$.

Step 2:
$\frac{2}{15} \div \frac{1}{5} = \frac{2}{15} \cdot 5 = \frac{2}{3}$.
$\frac{1}{8} \div \frac{1}{5} = \frac{1}{8} \cdot 5 = \frac{5}{8}$.

·		5/6	
1/5	2/15	1/6	1/8
	1/2		

·	2/3	5/6	5/8
1/5	2/15	1/6	1/8
	1/2		

Step 3:
$\frac{1}{2} \div \frac{2}{3} = \frac{1}{2} \cdot \frac{3}{2} = \frac{3}{4}$.

Final:
$\frac{3}{4} \cdot \frac{5}{6} = \frac{5}{8}$, and $\frac{3}{4} \cdot \frac{5}{8} = \frac{15}{32}$.

·	2/3	5/6	5/8
1/5	2/15	1/6	1/8
3/4	1/2		

·	2/3	5/6	5/8
1/5	2/15	1/6	1/8
3/4	1/2	5/8	15/32

189.

·	4/5	5/6	6/7
5/8	1/2	25/48	15/28
5/9	4/9	25/54	10/21
5/11	4/11	25/66	30/77

190.

·	2/3	7/8	2/5
2/5	4/15	7/20	4/25
1/3	2/9	7/24	2/15
3/4	1/2	21/32	3/10

191.

·	3/8	1/6	3/4
1/3	1/8	1/18	1/4
4/5	3/10	2/15	3/5
5/7	15/56	5/42	15/28

192.

·	3/5	1/2	2/3
1/6	1/10	1/12	1/9
4/9	4/15	2/9	8/27
3/4	9/20	3/8	1/2

193.

$\cdot$	$\frac{1}{2}$	$\frac{1}{3}$	$\frac{1}{4}$
$\frac{2}{5}$	$\frac{1}{5}$	$\frac{2}{15}$	$\frac{1}{10}$
$\frac{3}{8}$	$\frac{3}{16}$	$\frac{1}{8}$	$\frac{3}{32}$
$\frac{4}{9}$	$\frac{2}{9}$	$\frac{4}{27}$	$\frac{1}{9}$

194.

$\cdot$	$\frac{1}{2}$	$\frac{2}{3}$	$\frac{3}{4}$
$\frac{2}{3}$	$\frac{1}{3}$	$\frac{4}{9}$	$\frac{1}{2}$
$\frac{1}{4}$	$\frac{1}{8}$	$\frac{1}{6}$	$\frac{3}{16}$
$\frac{5}{6}$	$\frac{5}{12}$	$\frac{5}{9}$	$\frac{5}{8}$

195. To find the number of pieces, we divide the total length of the rope by the length of the pieces:

$$8 \div \frac{2}{3} = 8 \cdot \frac{3}{2} = 12.$$

So, Elle cut **12 pieces** with length $\frac{2}{3}$ inches.

196. A regular pentagon has five sides of equal length. So, the side length of a regular pentagon with perimeter $4\frac{5}{8}$ cm is

$$4\frac{5}{8} \div 5 = \frac{37}{8} \div 5 = \frac{37}{8} \cdot \frac{1}{5} = \frac{37}{40} \text{ cm}.$$

197. To find the width of the rectangle, we divide its area by its height:

$$5\frac{1}{3} \div \frac{4}{5} = \frac{16}{3} \div \frac{4}{5} = \frac{16}{3} \cdot \frac{5}{4} = \frac{20}{3}.$$

So, the rectangle is $\frac{20}{3} = 6\frac{2}{3}$ units wide.

198. To find the number of ounces in one serving, we divide the number of ounces in the can by the number of servings it holds:

$$22\frac{1}{2} \div 2\frac{1}{2} = \frac{45}{2} \div \frac{5}{2} = \frac{45}{2} \cdot \frac{2}{5} = 9.$$

So, each serving is **9 ounces** of soup.

199. To make just 24 cupcakes, Rob will make $\frac{24}{36} = \frac{2}{3}$ of the original recipe. So, he needs $\frac{2}{3}$ the amount of butter.

$\frac{2}{3}$ of $\frac{3}{4}$ is $\frac{2}{3} \cdot \frac{3}{4} = \frac{1}{2}$. So, Rob needs **$\frac{1}{2}$ of a cup** of butter.

200. After Grogg eats $\frac{1}{3}$ of the watermelon, $1 - \frac{1}{3} = \frac{2}{3}$ of the watermelon is left.

Two thirds of the watermelon weighs $7\frac{3}{4}$ pounds, so *one* third of the watermelon weighs

$$7\frac{3}{4} \div 2 = \frac{31}{4} \div 2 = \frac{31}{4} \cdot \frac{1}{2} = \frac{31}{8} = 3\frac{7}{8} \text{ pounds}.$$

Therefore, the whole watermelon originally weighed $\frac{31}{8} \cdot 3 = \frac{93}{8} = 11\frac{5}{8}$ **pounds**.

— *or* —

We use the approach discussed above to find the weight of one third of the watermelon, $3\frac{7}{8}$ pounds. The whole watermelon originally weighed

$$7\frac{3}{4} + 3\frac{7}{8} = 7\frac{6}{8} + 3\frac{7}{8} = 10\frac{13}{8} = 11\frac{5}{8} = \frac{93}{8} \text{ \textbf{pounds}}.$$

— *or* —

We think about the same problem with whole numbers.

If 4 watermelons weigh 12 pounds, then each watermelon weighs $12 \div 4 = 3$ pounds.

So, to find the total weight of one watermelon, we divide the weight by the number of watermelons.

Since $\frac{2}{3}$ of a watermelon weighs $7\frac{3}{4}$ pounds, one watermelon weighs

$$7\frac{3}{4} \div \frac{2}{3} = \frac{31}{4} \div \frac{2}{3} = \frac{31}{4} \cdot \frac{3}{2} = \frac{93}{8} = 11\frac{5}{8} \text{ \textbf{pounds}}.$$

201. We cancel as much as possible before multiplying:

$$\frac{\cancel{2}}{\cancel{7}} \cdot \frac{\cancel{13}}{3} \cdot \frac{\cancel{7}}{\cancel{17}} \cdot \frac{\cancel{11}}{\cancel{2}} \cdot \frac{\cancel{5}}{\cancel{11}} \cdot \frac{\cancel{17}}{\cancel{5}} \cdot \frac{19}{\cancel{13}} = \frac{19}{3} = 6\frac{1}{3}.$$

202. We simplify each expression grouped in parentheses, then divide:

$$\left(\frac{1}{\cancel{2}} \cdot \frac{\cancel{2}}{3} \cdot \frac{\cancel{3}}{4}\right) \div \left(\frac{3}{\cancel{4}_{2}} \cdot \frac{\cancel{2}}{5} \cdot \frac{\cancel{2}}{3}\right) = \frac{1}{4} \div \frac{1}{5} = \frac{1}{4} \cdot 5 = \frac{5}{4} = 1\frac{1}{4}.$$

203. We cancel as much as possible before evaluating:

$$\frac{\cancel{2} \cdot 3 \cdot 4 \cdot 5 \cdot \cancel{6} \cdot \cancel{7} \cdot \cancel{8}}{\underset{2}{\cancel{4}} \cdot \underset{2}{\cancel{6}} \cdot \underset{2}{\cancel{8}} \cdot \underset{2}{\cancel{10}} \cdot \underset{2}{\cancel{12}} \cdot \underset{2}{\cancel{14}} \cdot \underset{2}{\cancel{16}}} = \frac{1}{2^7} = \frac{1}{\mathbf{128}}.$$

204. Multiplication is commutative and associative. So, we first group each fraction with a whole number that shares some factors with the fraction's denominator:

$$\frac{2}{3} \cdot 11 \cdot \frac{5}{13} \cdot 19 \cdot \frac{2}{77} \cdot 26 \cdot \frac{1}{95} \cdot 27$$
$$= \left(\frac{2}{3} \cdot 27\right) \cdot \left(\frac{5}{13} \cdot 26\right) \cdot \left(\frac{2}{77} \cdot 11\right) \cdot \left(\frac{1}{95} \cdot 19\right)$$

Then, we rewrite each whole number as an equivalent fraction with denominator 1 and cancel:

$$\left(\frac{2}{3} \cdot \frac{27}{1}\right) \cdot \left(\frac{5}{13} \cdot \frac{26}{1}\right) \cdot \left(\frac{2}{77} \cdot \frac{11}{1}\right) \cdot \left(\frac{1}{95} \cdot \frac{19}{1}\right)$$
$$= \left(\frac{2}{\cancel{3}_1} \cdot \frac{\cancel{27}^{9}}{1}\right) \cdot \left(\frac{5}{\cancel{13}_1} \cdot \frac{\cancel{26}^{2}}{1}\right) \cdot \left(\frac{2}{\cancel{77}_7} \cdot \frac{\cancel{11}^{1}}{1}\right) \cdot \left(\frac{1}{\cancel{95}_5} \cdot \frac{\cancel{19}^{1}}{1}\right)$$

So, this expression simplifies to

$$\frac{2 \cdot 9 \cdot 5 \cdot 2 \cdot 2}{7 \cdot 5} = \frac{2 \cdot 9 \cdot \cancel{5} \cdot 2 \cdot 2}{7 \cdot \cancel{5}} = \frac{72}{7} = 10\frac{2}{7}.$$

Note that it isn't necessary to rewrite each whole number as a fraction. When factors in the denominator of a fraction cancel with a whole number, we usually write our work as shown:

$$\left(\frac{2}{\cancel{3}_1} \cdot \cancel{27}^{9}\right) \cdot \left(\frac{5}{\cancel{13}_1} \cdot \cancel{26}^{2}\right) \cdot \left(\frac{2}{\cancel{77}_7} \cdot \cancel{11}^{1}\right) \cdot \left(\frac{1}{\cancel{95}_5} \cdot \cancel{19}^{1}\right) = \frac{2 \cdot 9 \cdot \cancel{5} \cdot 2 \cdot 2}{7 \cdot \cancel{5}} = \frac{72}{7}.$$

205. We notice that $\frac{5}{13}$ appears in every grouped product, and all other fractions have the same denominator, 21.

So, we factor $\frac{5}{13}$ out of every expression in parentheses, then simplify.

$$\left(\frac{5}{13} \cdot \frac{8}{21}\right) + \left(\frac{5}{13} \cdot \frac{13}{21}\right) + \left(\frac{5}{13} \cdot \frac{17}{21}\right) + \left(\frac{5}{13} \cdot \frac{4}{21}\right)$$
$$= \frac{5}{13} \cdot \left(\frac{8}{21} + \frac{13}{21} + \frac{17}{21} + \frac{4}{21}\right)$$
$$= \frac{5}{13} \cdot \frac{42}{21}$$
$$= \frac{5}{13} \cdot 2$$
$$= \frac{10}{13}.$$

206. We have $\frac{99!}{101!} = \frac{99 \cdot 98 \cdot 97 \cdot \cdots \cdot 3 \cdot 2 \cdot 1}{101 \cdot 100 \cdot 99 \cdot \cdots \cdot 3 \cdot 2 \cdot 1}.$

Every number that appears in the numerator also appears in the denominator. So, we cancel all duplicated factors in the numerator and denominator.

$$\frac{\cancel{99} \cdot \cancel{98} \cdot \cancel{97} \cdot \cdots \cdot \cancel{3} \cdot \cancel{2} \cdot 1}{101 \cdot 100 \cdot \cancel{99} \cdot \cdots \cdot \cancel{3} \cdot \cancel{2} \cdot 1} = \frac{1}{101 \cdot 100}.$$

$101 \cdot 100 = 10{,}100$, so $\frac{99!}{101!}$ simplifies to $\frac{1}{\mathbf{10{,}100}}$.

207. The perimeter of the field is $120+50+120+50=340$ yards. It takes $6\frac{1}{4}$ cans to paint 340 yards of line. To find the number of yards that one full can will paint, we divide the total yards painted by the number of cans used:

$$340 \div 6\frac{1}{4} = 340 \div \frac{25}{4}$$
$$= 340 \cdot \frac{4}{25}$$
$$= \overset{68}{340} \cdot \frac{4}{\underset{5}{25}}$$
$$= \frac{68 \cdot 4}{5}$$
$$= \frac{272}{5}.$$

So, one full can will paint $\frac{272}{5} = \mathbf{54\frac{2}{5}}$ **yards of line**.

208. To find the median, we first put the list in order from least to greatest:

$$\frac{1}{2} < \frac{4}{7} < \frac{3}{4} < \frac{5}{6}.$$

The median of these four numbers is the number halfway between $\frac{4}{7}$ and $\frac{3}{4}$.
The number halfway between $\frac{4}{7}$ and $\frac{3}{4}$ is

$$\left(\frac{4}{7} + \frac{3}{4}\right) \div 2 = \left(\frac{16}{28} + \frac{21}{28}\right) \div 2$$
$$= \frac{37}{28} \div 2$$
$$= \frac{37}{28} \cdot \frac{1}{2}$$
$$= \frac{37}{56}.$$

So, the median of the four numbers is $\frac{\mathbf{37}}{\mathbf{56}}$.

209. The reciprocal of $\frac{100}{101}$ is $\frac{101}{100}$. To subtract, we use a common denominator of $100 \cdot 101 = 10{,}100$.

$$\frac{101}{100} - \frac{100}{101} = \frac{10{,}201}{10{,}100} - \frac{10{,}000}{10{,}100} = \frac{201}{10{,}100}.$$

So, $\frac{100}{101}$ is smaller than its reciprocal by $\frac{\mathbf{201}}{\mathbf{10{,}100}}$.

210. $\frac{1}{2}$ of $\frac{2}{3}$ of $\frac{3}{4}$ of Julian's age is $\frac{1}{2} \cdot \frac{2}{3} \cdot \frac{3}{4} = \frac{1}{4}$ of his age. Since $\frac{1}{4}$ of Julian's age is 9 years, Julian is $9 \cdot 4 = \mathbf{36}$ **years old**.

211. a. The largest unit fraction less than 1 is $\frac{1}{2}$.
$1 - \frac{1}{2} = \frac{1}{2}$. So, we have
$$1 = \frac{1}{2} + \frac{1}{2}.$$

Next, the largest unit fraction less than $\frac{1}{2}$ is $\frac{1}{3}$.
$\frac{1}{2} - \frac{1}{3} = \frac{3}{6} - \frac{2}{6} = \frac{1}{6}$, and $\frac{1}{6}$ is a unit fraction.
So, we can write 1 as the sum of three distinct unit fractions as shown:
$$1 = \frac{1}{2} + \frac{1}{3} + \frac{1}{6}.$$

Since $\frac{1}{3} + \frac{1}{4} + \frac{1}{5} = \frac{20}{60} + \frac{15}{60} + \frac{12}{60} = \frac{47}{60}$ *is less than 1, we cannot write 1 as the sum of any other three distinct unit fractions.*

b. In part (a), we learned that $1 = \frac{1}{2} + \frac{1}{3} + \frac{1}{6}$. We can always multiply both sides of an equation by the same number to get another true equation.

Multiplying both sides of $1 = \frac{1}{2} + \frac{1}{3} + \frac{1}{6}$ by $\frac{1}{m}$ gives

$$\frac{1}{m} \cdot 1 = \frac{1}{m} \cdot \left(\frac{1}{2} + \frac{1}{3} + \frac{1}{6}\right).$$

Distributing and simplifying, we have

$$\frac{1}{m} = \left(\frac{1}{m} \cdot \frac{1}{2}\right) + \left(\frac{1}{m} \cdot \frac{1}{3}\right) + \left(\frac{1}{m} \cdot \frac{1}{6}\right)$$
$$= \frac{1 \cdot 1}{m \cdot 2} + \frac{1 \cdot 1}{m \cdot 3} + \frac{1 \cdot 1}{m \cdot 6}$$
$$= \frac{1}{2m} + \frac{1}{3m} + \frac{1}{6m}.$$

c. In part b, we found $\frac{1}{m} = \frac{1}{2m} + \frac{1}{3m} + \frac{1}{6m}$.
So, when we replace m with 7, we get

$$\frac{1}{7} = \frac{1}{2 \cdot 7} + \frac{1}{3 \cdot 7} + \frac{1}{6 \cdot 7}$$
$$= \frac{1}{14} + \frac{1}{21} + \frac{1}{42}.$$

Check: $\frac{1}{14} + \frac{1}{21} + \frac{1}{42} = \frac{3}{42} + \frac{2}{42} + \frac{1}{42} = \frac{6}{42} = \frac{1}{7}$. ✔

In fact, there are 36 different sums of three distinct unit fractions that equal $\frac{1}{7}$. For example, if you used the method discussed on Page 73, then you would have instead found

$$\frac{1}{7} = \frac{1}{8} + \frac{1}{57} + \frac{1}{3{,}192}.$$

If you found a different sum of three unit fractions, be sure to check that the sum is equal to $\frac{1}{7}$.

212. Since the average of these three numbers is 1, their sum is $1 \cdot 3 = 3$. So, we have

$$\frac{3}{4} + \frac{4}{3} + n = 3.$$

$\frac{3}{4} + \frac{4}{3} = \frac{9}{12} + \frac{16}{12} = \frac{25}{12}$, and we have

$$\frac{25}{12} + n = 3.$$

Subtracting $\frac{25}{12}$ from both sides of the equation gives us

$$n = 3 - \frac{25}{12}$$
$$= \frac{36}{12} - \frac{25}{12}$$
$$= \frac{\mathbf{11}}{\mathbf{12}}.$$

213. The reciprocal of $\frac{a}{b}$ is $\frac{b}{a}$.
$\frac{a}{b} \div \frac{b}{a}$ is the same as $\frac{a}{b} \cdot \frac{a}{b}$.
So, the result of dividing $\frac{a}{b}$ by its reciprocal is $\frac{a}{b} \cdot \frac{a}{b} = \frac{a^2}{b^2}$.

$$1 \qquad \frac{a}{b} \qquad \frac{b}{a} \qquad \frac{2a}{2b} \qquad \boxed{\frac{a^2}{b^2}} \qquad \frac{b^2}{a^2}$$

214. We begin by converting $\frac{1}{2}, \frac{1}{3}, \frac{1}{4}$, and $\frac{1}{5}$ into fractions with a common denominator. The LCM of 2, 3, 4, and 5 is 60, so we have

$$\frac{1}{2} = \frac{30}{60}, \quad \frac{1}{3} = \frac{20}{60}, \quad \frac{1}{4} = \frac{15}{60}, \text{ and } \frac{1}{5} = \frac{12}{60}.$$

So, any weight Grogg can make with these weights can be written as a number of sixtieths.

The largest fraction that is less than 1 and has a denominator of 60 is $\frac{59}{60}$.

To check that Grogg can make $\frac{59}{60}$ grams with the given weights $\left(\frac{30}{60}, \frac{20}{60}, \frac{15}{60}, \text{and } \frac{12}{60}\right)$, we try to write 59 as the sum of some combination of 30's, 20's, 15's, and 12's.

To get a units digit of 9, we must add one 15 and two 12's: $15 + 12 + 12 = 39$.

Then, we add one 20 to get a total of 59:

$$15 + 12 + 12 + 20 = 59.$$

So, Grogg can make a weight of $\frac{59}{60}$ grams using one $\frac{1}{4}$-gram weight, two $\frac{1}{5}$-gram weights, and one $\frac{1}{3}$-gram weight:

$$\frac{1}{4} + \frac{1}{5} + \frac{1}{5} + \frac{1}{3} = \frac{15}{60} + \frac{12}{60} + \frac{12}{60} + \frac{20}{60} = \frac{59}{60}. \checkmark$$

Therefore, $\frac{59}{60}$ **grams** is the largest weight less than 1 gram that Grogg can make.

215. *Three* fourths of the distance uses $\frac{2}{3}$ of a tank of gas, so *one* fourth of the distance uses $\frac{2}{3} \div 3 = \frac{2}{3} \cdot \frac{1}{3} = \frac{2}{9}$ of a tank.

If *one* fourth of the distance uses $\frac{2}{9}$ of a tank of gas, then *the whole distance* uses $\frac{2}{9} \cdot 4 = \frac{8}{9}$ of a tank.

— *or* —

We think about the same problem with whole numbers.

If 6 tanks of gas are used to drive 3 trips to Grandma's, then $6 \div 3 = 2$ tanks of gas are used to drive 1 trip to Grandma's.

So, to find the number of tanks of gas used to make one trip, we divide the number of tanks by the number of trips.

Since $\frac{2}{3}$ of a tank of gas is used to drive $\frac{3}{4}$ of the trip, $\frac{2}{3} \div \frac{3}{4} = \frac{2}{3} \cdot \frac{4}{3} = \frac{8}{9}$ of a tank is used to make one trip.

216. Winnie poured out $\frac{1}{3} - \frac{1}{5} = \frac{5}{15} - \frac{3}{15} = \frac{2}{15}$ of the pitcher.

Therefore, $\frac{2}{15}$ of the pitcher is 8 oz.

Two fifteenths of the pitcher is 8 oz, so *one* fifteenth of the pitcher is $8 \div 2 = 4$ oz.

$\frac{1}{5} = \frac{3}{15}$ of the pitcher is filled with juice after Winnie pours.

One fifteenth of the pitcher is 4 ounces, so *three* fifteenths of the pitcher is $3 \cdot 4 = 12$ ounces.

So, there are **12 ounces** of juice left in the pitcher.

Check: If 12 ounces of juice are left in the pitcher, then there were $12 + 8 = 20$ ounces of juice in the pitcher before Winnie poured out 8 ounces.

Also, if 12 ounces of juice is $\frac{1}{5}$ of the pitcher, then the pitcher holds a total of $12 \cdot 5 = 60$ ounces. So, before Winnie poured out the juice, those 20 ounces of juice are $\frac{20}{60} = \frac{1}{3}$ of the pitcher. $\checkmark$

For additional books, printables, and more, visit
www.BeastAcademy.com